COMING TO OUR SENSES

COMING TO OUR SENSES

A Boy Who Learned to See,
a Girl Who Learned to Hear, and
How We All Discover the World

SUSAN R. BARRY

BASIC BOOKS
NEW YORK

Cover design by Chin-Yee Lai
Cover image © CSA Images via Getty Images
Cover copyright © 2021 Hachette Book Group, Inc.

Basic Books
Hachette Book Group
1290 Avenue of the Americas, New York, NY 10104
www.basicbooks.com

Printed in the United States of America
First Edition: June 2021

Published by Basic Books, an imprint of Perseus Books, LLC, a subsidiary of Hachette Book Group, Inc. The Basic Books name and logo is a trademark of the Hachette Book Group.

The Hachette Speakers Bureau provides a wide range of authors for speaking events. To find out more, go to www.hachettespeakersbureau.com or call (866) 376-6591.

The publisher is not responsible for websites (or their content) that are not owned by the publisher.

Additional copyright/credits information is on page 217.

Print book interior design by Linda Mark

Library of Congress Cataloging-in-Publication Data
Names: Barry, Susan R., author.
Title: Coming to our senses : a boy who learned to see, a girl who learned to hear, and how we all discover the world / Susan R. Barry.
Description: First edition. | New York : Basic Books, 2021. | Includes bibliographical references and index.
Identifiers: LCCN 2020036615 | ISBN 9781541675155 (hardcover) | ISBN 9781541675162 (ebook)
Subjects: LCSH: People with disabilities—Rehabilitation—United States. | Blind—Rehabilitation—United States. | Hearing impaired—Rehabilitation—United States. | People with disabilities—United States—Biography. | Blind—United States—Biography. | Hearing impaired—United States—Biography. | Senses and sensation—Social aspects—United States.
Classification: LCC HV3023.A3 B37 2021 | DDC 362.4/1092 [B]—dc23
LC record available at https://lccn.loc.gov/2020036615

ISBNs: 978-1-5416-7515-5 (hardcover); 978-1-5416-7516-2 (ebook)

LSC-C

Printing 1, 2021

For
Cindy Lansford
and
Najma Gulamali Mussa

The key to the future of the world is finding the optimistic stories and letting them be known.

—Pete Seeger, American folksinger and songwriter

Contents

Introduction

Blessing or Curse?

RICHARD GREGORY AND JEAN WALLACE WERE EXCITED TO MEET Sidney Bradford. The fifty-two-year-old man had just undergone cornea operations that allowed him to see for the first time.[1] What would this be like? Would he be filled with gratitude and joy? Would he spring up from his hospital bed, look around, and find a fascinating new world spread out before him? Indeed, SB (as Gregory and Wallace referred to him) did experience an initial period of excitement and curiosity. A cheerful and extroverted man, he took great delight in watching the traffic go by and distinguishing cars from trucks. But over the following months, his mood shifted. Even with his new sight, he couldn't read print or drive a car. He had led a full life as a blind man and was perfectly healthy when he underwent the operations that gave him sight, but over the year and a half following the surgeries, he grew increasingly depressed, his health deteriorated, and he died.

Virgil, the subject of Oliver Sacks's essay "To See and Not See," fared no better.[2] Blinded by cataracts in infancy, Virgil underwent operations to remove them in midlife. Now, he could see but could hardly make sense of what he saw. Within a year of the operations, he became gravely ill and lost what vision he had.

In *Wired for Sound*, Beverly Biderman describes her initial despair upon first receiving a cochlear implant.[3] Biderman began to lose her hearing as a child and had been deaf for more than thirty years before she received her device. After experiencing sounds again, she had an "overwhelming feeling of things falling apart." She could not go back to being deaf, yet found her current situation intolerable. Completely out of balance, she wrote, "I felt quite simply that I wanted to die."

WHY IS THIS SO HARD? WHY WOULD A BLIND PERSON NOT EMbrace sight and a deaf person not jump at the chance to hear? I found this question particularly intriguing because at age forty-eight, I experienced a dramatic improvement in my vision, a change that repeatedly brought me moments of childlike glee.[4] Cross-eyed from early infancy, I had seen the world primarily through one eye. Then, in midlife, I learned, through a program of vision therapy, to use my eyes together. With each glance, everything I saw took on a new look. I could see the volume and 3D shape of the empty space between things. Tree branches reached out toward me; light fixtures floated. A visit to the produce section of the supermarket, with all its colors and 3D shapes, could send me into a sort of ecstasy. If I was so enchanted by my new stereo views, why wouldn't a blind person, enabled to see for the first time, be overwhelmed with joy?

There's a difference between seeing in a new way and seeing for the first time. My new stereo views brought me great pleasure because they didn't disrupt, but rather confirmed, my overall worldview. I could always order objects in depth using cues seen by just one eye. For example, an object in the foreground blocked my view of objects located behind it. One eye gave me depth order but with a compressed sense of space. When I gained stereovision, space inflated. I could see, not just infer, the pockets of space between a given object and those behind it. My new 3D views made sense.

The same is not true for an adult or older child who gains sight or hearing for the first time. To the newly sighted, the ease with which most of us use our eyes to interpret the world is staggering. Following a single glance, we grasp the gist of the scene before us.[5] But where we perceive a three-dimensional landscape full of objects and people, a newly sighted adult sees a hodgepodge of lines and patches of colors appearing on one flat plane. One twenty-five-year-old woman described her first views in this way: "I see an ensemble of light and shadow, lines of different length, round and square things, generally like a mosaic of changeable sensations, that astonish me, and whose meaning I do not understand."[6]

As Albert Bregman points out in his book *Auditory Scene Analysis*, sensory input does not come out of nowhere.[7] We don't see greenness in the absence of something that is green. We don't hear loudness in the absence of a sound-causing event. Colors, textures, and contours, squeaks, bangs, and voices belong to something or someone. An adult gaining sight or hearing for the first time is barraged with new sensations that do not seem to belong anywhere. They are disembodied and meaningless. Indeed, ophthalmologist Alberto Valvo quoted the words of one sight-recovery patient: "This is too long and unhappy a road, leading one into a strange world. I may even have been happier before this: now I seem weak, and I am often seized by a great feeling of fatigue."[8]

When I look at the scene in Figure I.1, I can easily see three separate objects: a cup, a spoon, and a bowl. Although the cup partially blocks my view of the spoon, I recognize the two visible parts of the spoon as belonging to the same object. But to an adult or older child just gaining sight, the photograph may appear as a design on one flat plane so that the two parts of the spoon belong to two separate entities. Shadows in the scene only add to the confusion. Yet the

FIGURE I.1. The cup partially blocks our view of the spoon's handle, but we recognize that the handle continues behind the cup.

photograph presents a far less complex view than that of an ordinary kitchen, yard, street, or landscape.

The same problem of "belongingness" occurs with audition. As I type on this hot, humid day, I recognize the patter of rain through my open window, the whir of the fan by my desk, the tap of my keystrokes, and my husband's voice. I turn on the radio and hear an orchestra playing. Although the violin and flute are simultaneously playing the same notes, I have no problem recognizing which sounds belong to each instrument. Each of these separate sounds (the rain, the fan, the keystrokes, the violin and flute) is made up of a combination of sound waves arriving at my ears at the same time; yet I automatically sort them all into their proper sound sources. I know which sound waves belong to the rain and which to the flute. An adult hearing for the first time, however, perceives an unintelligible cacophony. Sounds lack meaning, are disembodied, and are hard to localize. A newly acquired sense in adulthood does not add to one's understanding of the world; it confounds it.

> For I am now, after forty years of what we will term silence, so accommodated to it (like a hermit-crab to its shell) that were the faculty of hearing restored to me tomorrow it would appear an

> affliction rather than a benefit. I do not mean that I find deafness desirable but that in the course of time the disability has been assimilated to the extent that it is now an integral condition of existence, like the use of a hand. By the same token the restoration of my hearing, or the loss of my deafness, whichever is the right way of putting it, would be like having that hand cut off.[9]

These are the words of the poet David Wright, who, deafened at the age of seven, pondered what it would be like to hear again. For him, this was a hypothetical exercise—no treatments were available at the time of his writing to restore his hearing, but a few years later, in 1972, the first cochlear implants were introduced. Many deaf people objected to them.[10] As members of the Deaf community, with their own language, Sign, and their own culture, they understood better than most hearing people the enormous adjustment, both personal and social, that the acquisition of hearing would demand.

To ask the blind or deaf to acquire a new sense past childhood is to ask them to reshape their identity. They may have functioned independently quite well before but now find themselves as vulnerable as a young child. With their new sight, they see but cannot recognize a flight of stairs or a loved one's face. With their new hearing, they sense but don't understand a word that is spoken. After gaining sight at age thirty, John Carruth, who had once navigated easily in the dark, lost all confidence and moved in the dark with caution.[11] When SB was blind, he crossed streets with confidence, but after first gaining sight, he found the traffic frightening and wouldn't cross streets by himself.[12] And this feeling of helplessness in the newly sighted or newly hearing intensifies with the knowledge that vision and hearing provide others with much more detail about them than they previously realized. Indeed, Valvo reported one patient who, after gaining functional vision, continued to use his dark glasses and cane when going out. This man felt admired when walking as a blind person; he did not want people to pity him when they saw him walking more hesitantly using his sight.[13]

Sight or hearing, if newly acquired, can plunge a person into spatial confusion. The newly sighted do not have the experience to know visually how far objects are from them.[14] When SB first looked out a window that was thirty to forty feet above the ground, he thought he could safely lower himself down by his hands.[15] Without vision, distance and space must be judged by other means. As John Hull, blinded in midlife wrote, "Space is reduced to one's own body, and the position of the body is known not by what objects have been passed but by how long it has been in motion. Position is thus measured by time."[16] TG, a sight-recovery patient described by Valvo, echoed Hull's words: "Before the operation, I had a completely different idea of space . . . when I was blind I would refer only to the time necessary before arriving at a given point. After the operation I had to coordinate both vision and the time necessary for the distance involved, and I could not manage that."[17] Not only do the newly sighted have to develop a new concept of space and distance, but, with their vision, they must develop a new perceptual style. We explore the world with our hands and ears in a sequential fashion, touching one point after another or hearing a sequence of sounds, but, with our eyes, we see many things together in one moment.

Although sight allows us to sense things at a distance, we cannot see things behind obstacles, around corners, or in the dark. But we can hear them. The way sounds bounce off objects and walls helps us to know, even without sight, whether we are in a small, enclosed room or a large, open space. The perceptual world of a person who is deaf is both organized and limited by what he or she can see. A deaf person who receives a cochlear implant struggles not only to recognize the sounds but also to know where they come from. Sounds and their echoes appear out of the blue, disrupting their understanding of where they and other things are in space.

While most of us cannot imagine what it is like to gain a new sense, many of us can remember how disruptive it was to move to a new home. Even if the new house provided a better place to live, the prospect of leaving the old neighborhood, so constant and familiar, was frightening. Things at the new place were not where they used to be, so daily habits and movements had to change. Such adaptations require brain reorganization, and as scientist and writer I. Rosenfield points out in his book *The Invention of Memory*, this reorganization can lead to anxiety and depression.[18] The acquisition of a new sense requires leaving a familiar perceptual world and reckoning with new relationships with almost everything around you. As we shall see, such a move necessitates much greater reorganization of brain circuitry than occurs with relocating to a new house, so the possibility of increased anxiety and depression is very real.

AT FIRST BLUSH, VISION AND HEARING MAY SEEM PURELY MECHANical processes. Photons hit the light-sensing pigments in the retina of the eye, triggering a cascade of electrical and chemical events that signal to the brain about light, color, and motion. Sound waves of different frequencies vibrate different parts of the cochlea in the inner ear, giving us our sense of pitch. Yet these events tell only part of the story. Even if we all possessed identical sensory structures, we would each perceive a different and very personal version of the world, a version built upon our experiences, needs, and desires.

John Hull wrote that blindness was "a state" "like the state of being young, or being old, of being male or female; it is one of the orders of human being . . . One human order finds it difficult to understand another."[19] In a similar way, adults or older children first gaining sight or hearing have lived in a perceptual world so different from most of ours that their descriptions of their first sights or sounds are hard to imagine. They remind us that our perception is shaped not just by our eyes and ears but by the whole of our experience.

I was astonished, when I gained stereovision, by how different the world looked. Since I inhabited the same world as everyone else, I assumed that I saw it in much the same way. After all, I could identify the objects around me and talk about them with others. A tree was a tree, but, after I gained stereovision, it took on a whole new look. Its canopy of leaves no longer appeared flat, as in a child's drawing. Instead, I saw layers upon layers of leaves and branches. When I looked in the mirror, I no longer saw my reflection on the surface of the glass; instead, I saw it behind the mirror in the reflected space. The most striking thing about this change occurred when I momentarily closed one eye: I didn't revert to my old stereoblind way of seeing but still saw my reflection as behind the glass. My view through one eye was now influenced by my experience of seeing through two. When I described my new sights to people who had always had stereovision, they were baffled, having never imagined that someone would see their reflection not behind but on the surface of the mirror. Yet, when I mentioned my new vision to people who had always been stereoblind, they were mystified that my reflection could be located anywhere else but on the plane of the glass. There was a perceptual gap between those who always saw and those who never saw with stereovision, one that couldn't be entirely bridged. In the same way, any sighted or hearing person can never fully imagine what it would be like to see or hear for the first time.

Indeed, we begin to mold our perceptual world from birth. Newborn babies may appear helpless, but they are not passive recipients of the stimuli around them. At or shortly after birth, infants can recognize their mother's voice, and within days they recognize her face as well. During the first year, they become particularly sensitive to the sounds of their native language and the kinds of faces they habitually see. They also possess an irrepressible urge to explore and experiment. Starting at about four months old, when they can reach for objects, they can't resist squeezing, shaking, or dropping them or smashing two things together. In this way, they teach themselves about the properties of those objects and their three-dimensional shapes. Although we all employ the same general mechanisms and brain areas to organize and process sensory information, each child's perceptual system develops in its own unique way to match the people and objects in their particular environment and to pick up the information most important to them.[20]

In describing the private nature of our perceptions, Oliver Sacks wrote, "Every perception, every scene is shaped by us, whether we intend it or know it, or not. We are the directors of the film we are making—but we are its subjects too: every frame, every moment, is us, ours."[21] Just as the movie cameraman and sound engineer direct the cameras and microphones to the action that they want the audience to follow, we move our body, head, and eyes to select what we see and hear. Our sharpest, most acute vision is located in the fovea, the central part of the eye's retina. So to see an object in detail, we must look directly at it. We turn our heads to face the source of an interesting sound in order to see it best. As we scan a scene, we move our eyes from one point to another, pausing to gaze at or fixate on important features. Studies that monitor people's eye movements while they view different scenes indicate that we don't all scan a scene in the same way.[22] The volume of raw stimuli out there can be overwhelming, so we all must select what to focus on and what to ignore. Where we direct our eyes, what we attend to, depends upon our prior knowledge of the surroundings, our past experiences and preferences, the task at hand, and what we predict will happen next.[23]

Sights and sounds are laden with private associations and emotions that, throughout life, influence what we attend to and what we perceive. One summer day, my ten-year-old son and I took a walk along a winding road near the Cape Cod shore. As we strolled, I mused about the birds and the trees, but my son hardly heard me. He pointed instead to the telephone poles, where he spied powerlines and voltage transformers and then explained to me how they all worked. We were walking along the same path and glancing in the same general direction but seeing very different things, filtering in what we recognized and what was important to us while ignoring the rest. For my son, the trees, which I found so beautiful, and for me, the powerlines, which he found so fascinating, were just background noise. But since that walk, I do notice powerlines and voltage transformers because I now know what they do and because they are associated with a happy memory. Perception shapes experience, and experience shapes perception. If the events of an ordinary day, like my walk with my son, alter perception, then the acquisition of a new sense leads to much more radical transformations, which will be different and personal for each individual.

Throughout life, we continue to tune our sensory systems to our particular environment, needs, and expertise. A mechanic looking at a car engine sees more with a single glance than most of us do. During a walk in the forest, we may all notice the same bird, but the birdwatcher extracts more information from what she sees. She has learned what characteristics and patterns (such as beak, feathers, flight behavior, bird calls) to attend to in order to identify a particular species. Psychologist Eleanor Gibson called learning to extract and select the most relevant features and patterns from all the information out there "perceptual learning."[24] Perceptual learning is different from the learning of facts, as may occur in school, or the learning of new motor skills, such as hitting a baseball. Although we've been acquiring knowledge through perceptual learning since infancy, we don't always know exactly how or what we've learned. How is it that we recognize, for example, both a Saint Bernard and a miniature dachshund as dogs? Can you explain all the information we use to make this determination? For adults just gaining sight, perceptual learning in the visual sphere has to start at a very basic level. The newly sighted can't identify what kind of bird they are seeing until they can first recognize the bird as a single unit separate from the tree branch it is perched on and as a visual category of animal distinct from all others.

Each year, when teaching introductory biology at Mount Holyoke College, I witnessed perceptual learning in my students. For example, I took those in my introductory biology class on a field trip around the Mount Holyoke campus. As we approached the lake, I stopped and asked them if they saw any new plants. "Do you mean those flowers?" one student asked pointing to a stand of late-blooming asters. When I shook my head no, another student pointed to some ferns.

"Not those either," I said. "Keep looking." Several of the students moved down to the very edge of the water but found nothing out of the ordinary. Finally, after some hints from me, one student asked, "Do you mean those green stems over there?"

"Those green stems" were horsetails, and that is what I wanted the students to see. I explained to them that horsetails were ancient, nonflowering plants that dominated the forests 500 million years ago. As we continued our walk, students kept pointing out the horsetails. Earlier, these plants had been invisible to them, but now that I had drawn their attention to them, they kept popping into focus. The world outside them had not altered, but they were extracting new information from it. Their personal perceptual world had changed.

How we attend to the world influences not only what we perceive but who we are. Those who are most aware of a person's facial expressions, body language, and vocal tones may be particularly sensitive to the thoughts and feelings of others. Those who keep track of the sun's position in the sky as they move from place to place may develop a good sense of direction. Our likes and dislikes and our ways of attending to and perceiving the world are interdependent and mutually reinforce each other. We hone our perceptual skills for the activities we enjoy most, and we prefer to engage in activities we perceive best. If we are captivated by the sound of the piano, we'll listen to and play it more, becoming more sophisticated listeners, which further increases our pleasure. Whether we are conscious of it or not, we all have our own set of perceptual biases, our own perceptual style, which both guides and limits what we sense and do. Perception is a personal act. If we want to become better observers and listeners, we must become aware of how we use our eyes and ears and attend to the world in new ways.

"Perception is not something that happens to us, or in us. It is something we do," philosopher Alva Noë writes.[25] We move our body, head, and eyes to look and listen, to take in information about the world. Since we direct what we see and hear, developing vision or hearing as an adult is an intensely active process. A new pair of eyes or ears won't lead to vision or hearing unless the owner of those new eyes or ears pays attention to what he is sensing and figures out its meaning.

In *Space and Sight*, Marius von Senden describes the experience of two boys, both about five years old, blinded from birth, whose sight was restored following cataract surgery.[26] To the astonishment of their surgeons, the boys didn't react to their new sight at all but continued to explore with their hands. Even after they were given instructions on how to use their vision, they ignored what they could see. Since what they saw lacked meaning, they didn't intentionally look at anything or assimilate their new sensations into their personal worldview.

Even after he gained sight, SB did not turn his eyes to the source of a noise or look at people's faces, and he rarely scanned his surroundings the way normally sighted people do. His primary means of exploring the world was through touch. "It would seem that the difficulty," Gregory and Wallace wrote, "is not so much in learning *per se* as in *changing perceptual habits and strategies from touching to seeing*."[27] Sacks, too, emphasized how difficult it is to change the way we habitually take in the world. When Virgil gained sight, he bought toy replicas of people, cars, and animals and tried to correlate the feel of these toys with the images of the real people, cars, and animals he could now see. On a trip to the zoo, Virgil initially thought the gorilla he saw looked just like a man. Only after he passed his hands expertly over a nearby statue of the animal could he see the difference. After watching Virgil at the zoo, Sacks wrote, "Exploring it [the gorilla statue] swiftly and minutely with his hands, he had an air of assurance that he had never shown when examining anything by sight. It came to me . . . how skillful and self-sufficient he had been as a blind man, how naturally and easily he had experienced his world with his hands, and how much we were now, so to speak, pushing him against the grain: demanding that he renounce all that came easily to him, that he sense the world in a way incredibly difficult for him, and alien."[28] To learn to see or hear past early childhood requires fundamental changes in perceptual habits and behavior, a major reordering of one's perceptual world, and intense exploration, experimentation, and analysis.

It's not surprising then that most congenitally deaf people who successfully learn to hear have received cochlear implants as infants or young children. And cases of sight recovery past childhood are extremely rare. Although there are reports of newly sighted adults who acclimated to their vision, von Senden, who reviewed sixty-six cases of sight recovery, concluded that the initial excitement of first seeing is almost always followed by a psychological crisis.[29] Experimental studies on laboratory animals add scientific weight to these pessimistic reports. Such experiments have suggested that sensory deprivation during a critical developmental period in early life leads to lifelong, irreversible sensory impairments. For example, covering one eye in an infant cat or monkey, though not in an adult, leads to changes in brain organization that favor input from the uncovered eye and results in a loss of binocular vision.[30] Until recently, therefore, few attempts were made to restore vision or hearing in congenitally blind or deaf people older than eight years. By age eight, the brain, it was thought, was no longer plastic enough to allow for the development of a new sense.

So I was greatly intrigued to meet in one year two people who acquired a new sense past early childhood, both of whom not only accommodated but embraced it. Liam McCoy, who was practically blind since infancy, gained sight following a series of bold operations at age fifteen. Zohra Damji was profoundly deaf until receiving a cochlear implant at the relatively late age of twelve. Indeed, her surgeon told her aunt that, had he known the length and degree of Zohra's deafness, he would not have performed the operation.

Liam and Zohra join a rare group of individuals who have recovered a new sense past early childhood and adapted to it. After her initial struggles, Beverly Biderman embraced her restored sense of hearing.[31] Von Senden and Valvo described several patients with happier outcomes;[32] the newly sighted Michael May, as described in *Crashing Through*, adapted to his vision;[33] and, as discussed in Chapter 6, many of the children and young adults treated in the sight-recovery program Project Prakash make good use of their new vision.[34] What allows some to find their way in a new and initially chaotic perceptual world, while others do not? There is no single answer to this question, for the outcome varies from person to person and depends upon the whole history and tenor of a person's life. Each of us perceives and adapts to the world in our own unique way.

I met Liam when he was twenty years old, five years after his surgeries, and Zohra, when she was twenty-two, ten years after her cochlear implant. While both described to me their childhood and the initial shock of first seeing and hearing, we did not try to reconstruct their day-by-day progress in adapting to their new sense; we explored instead how they perceive the world today. Only by learning of their childhoods and sharing a small part of their daily lives could I begin to understand how they reconstructed and reordered their perceptual worlds. So we discussed the challenges and successes they experienced when young, the support they received from family and doctors, their education, their goals, and the perceptual training and strategies they developed in order to find their way in a society that assumes that everyone can see and hear. We exchanged scores of emails; I visited their homes, met their families, and experienced with them the frustrations and pleasures of an ordinary day. Over the course of a decade, Liam and Zohra told me their stories, stories that reveal the personal and private nature of our perceptions and our power to transform and adapt them to the physical and social worlds we all share.

PART I

LIAM

To give back his sight to a congenitally blind patient is more the work of an educationist than that of a surgeon.

—F. Moreau, quoted in M. von Senden, *Space and Sight: The Perception of Space and Shape in the Congenitally Blind Before and After Operation* (Glencoe, IL: Free Press, 1960), 160.

– chapter 1 –

How Far Is Your Vision?

IN THE OPENING SCENE OF A BEST-SELLING NOVEL, THE CURATOR of a famous museum is shot and killed. His murderer, we later learn, is a white-haired man with albinism. For my friend, Liam McCoy, the selection of the murderer makes no sense. Liam has albinism and knows that most people with his condition could not have carried out the murder. Their vision is just too poor.

I first met Liam through his ophthalmologist, Dr. R. Lawrence Tychsen, who had invited me to give a talk to the Department of Ophthalmology at Washington University in St. Louis. Dr. T. (as Liam calls him) treats children with neurological impairments, often so severe that other doctors consider them too difficult to examine and treat. He was intrigued by my story of gaining stereovision as an adult and then told me about one of his patients with a remarkable vision-recovery story. "You must meet Liam," he said to me, and so, over many phone calls, emails, and visits, Liam told me his story.

IT WAS OBVIOUS TO THE OBSTETRICS NURSE THAT SOMETHING WAS different about Liam the moment his head crowned. Liam's hair was metallic silver, and blood vessels were plainly visible through his very light-colored skin. "Oh my God!" the nurse exclaimed as she rushed from the delivery room. Moments later, she returned with the doctor, who took one look at the newborn and hurried out too. When the doctor returned, Cindy, Liam's mom, now deeply concerned, asked what was wrong. "Oh, he's a towhead; he's a cotton-top," the doctor responded. But it would have been better had the doctor not been quoted the next day in the hospital newsletter as saying, "I've never seen a baby with such silver blonde hair!" Strangers kept stopping into the hospital room to peek at the new baby. Cindy took Liam home as soon as possible, just to get some rest.

Liam had a mohawk of silver hair running from the back to the front of his head. But when Cindy tried to take pictures of the silver strands, all the photos came out overexposed. Her baby's hair was just too pale to show up well in pictures. From the start, Cindy suspected that her child had albinism. She had known several people with this condition. By coincidence, at her previous workplace, the cafeteria had been staffed (cooks, waiters, and waitresses) entirely by people with albinism. So when Liam was just one week old, Cindy asked the pediatrician about her concerns. The pediatrician dismissed them. Liam's eyes were a pale blue. He had Nordic relatives on his father's side so, the doctor thought, he might favor them. In retrospect, his misdiagnosis is not surprising. Albinism, or a lack of pigment in the hair, eyes, and skin, is a rare condition, affecting only one in seventeen thousand people. And in contrast to incorrect and often cruel media portrayals, people with albinism do not have pink or red eyes. Like Liam, their eyes are blue, gray, or sometimes violet. So, it was not until Liam was seventeen months old that a genetics specialist confirmed that he had albinism.

Liam's blue eyes result from a lack of the pigment, melanin, in the iris, the colored part of the eye that controls the size and diameter of the pupil. While melanin makes our eyes green or brown, there are no pigments that make our eyes blue. The iris is made up of several tissue layers, and the blueness results from the way light is scattered by these layers. (A similar phenomenon makes the sky blue.) Not just individuals with albinism but all people with blue eyes lack melanin in the front of the iris. What distinguishes people with albinism from others with blue eyes is that they have little or no melanin in other parts of the iris and eye and, in many cases, in the skin and hair as well. Indeed, several gene mutations have now been discovered that result in albinism, and all of these mutations affect the synthesis of melanin throughout the body.

Many people think of people with albinism as pigmentless. But this is not the case. Melanin is only one class of the many pigments found in our bodies. Other pigments include hemoglobin, the oxygen-binding molecule in our blood, as well as rhodopsin and photopsin, the light-sensing pigments found in the rod and cone cells of our eyes. People with albinism have these other pigments; only melanin is reduced or lacking.

It must have been lonely for Cindy as a new mother. She could not take her baby for walks or to the park on a bright, sunny day. Like many babies with albinism, Liam was photophobic, extremely sensitive to bright light. Here, too, the lack of melanin is to blame. Melanin is found not only in the front of the iris but also in the back, where it prevents light from entering the eye. Because of melanin, light enters our eye through only one point, the pupil. In bright sunshine, we contract our pupil to reduce the amount of light that hits the retina, while, in the dark, we dilate our pupils to allow in more light. Without melanin in the back of the iris, people with albinism have a harder time regulating how much light enters the eye, and the extra light causes painful glare. So Cindy took Liam out only at night or at dawn or dusk. She put up special utility lights around her house for watching Liam after sunset and got the neighbor's permission to put Liam's toddler pool by their wall in the shade between the two houses.

When Liam was four months old, Cindy grew increasingly concerned about his vision. She had just started feeding him solid food. She would fill the spoon and then move it back and forth slowly in front of him. Even though Liam was hungry, he did not follow the spoon's movements with his eyes. So Cindy took Liam to the pediatrician and asked about his vision. The doctor walked around the room with a flashlight in order to see if Liam followed the light with his eyes. The doctor claimed that he could see just fine.

But Cindy had been trained as a speech pathologist working with blind and deaf children. She was not confident in the doctor's conclusions. To stimulate Liam's vision, Cindy clamped a light onto the changing table and kept it on at all times to give Liam something to see and orient to. She taped a Dairy Queen ad, full of red and black colors, to Liam's crib. She also noticed that Liam's eyes wandered, and this eye misalignment, or strabismus, did not improve with age. When Liam was old enough to stand and talk to his mom, his right eye would move up and out so that it seemed to Cindy that Liam was looking over her left shoulder. Liam remembers his eyes as uncontrollable. Like most people with albinism, he had nystagmus, an involuntary oscillating movement of the eyes. He could not willingly look at anything.

Despite problems with his vision, Liam developed motor skills faster than most babies. One moment, he managed to get up on all fours, and the next moment he was crawling. His balance was excellent. Once Cindy snapped a photo of Liam balancing on a moving rocking chair before ordering him to get down. Liam began to walk quite early, at seven to nine months, but he grabbed hard onto Cindy's fingers when he walked and would not let go. He needed Cindy for visual guidance, not balance. Up and down their home they marched until one day, when Liam was just over a year old, he saw a bright reflection on a filing cabinet two or three feet away. At that moment, Cindy was across the room folding clothes, but she spied Liam as he let go of a laundry hamper and walked off independently, without even a wobble, to the shiny cabinet to investigate. Liam's motor skills and liveliness remind me of the young Helen Keller as described by her teacher, Anne Sullivan.[1] Though Keller was blind and deaf, she spent her days running, jumping, spinning, swimming, and even climbing trees. She was, according to Sullivan, "graceful as a nymph."

At sixteen or seventeen months old, Liam developed a rash. Cindy took him to the pediatrician's office, but their normal doctor wasn't in, so they saw his partner instead. "Whom does he see for his eyes?" asked this doctor immediately upon entering the room. In contrast to their normal pediatrician, he noticed immediately that Liam's eyes didn't work together. Finally, Cindy had confirmation that something was wrong with Liam's vision. The doctor referred them to a pediatric ophthalmologist, whom they went to see soon afterward. But this visit proved disastrous. He examined Liam and abruptly announced, "He's blind, and there's nothing we can do about it. He can see the big *E* and that's all he'll ever see."

Later that same day, Cindy went shopping and took Liam with her. It was almost Christmastime. Cindy picked out some toys and hid them under other items in her shopping cart. When the cashier started to ring up a toy, Liam asked if it was for him. "Blind? My foot!" Cindy thought.

Our eyes are not fully formed at birth. Indeed, they may continue to develop up to age eight, and our vision takes even longer to mature.[2] If a baby could read, he still couldn't see and identify the letters on the standard Snellen eye chart some twenty feet away. A newborn's visual acuity is much poorer than that of an adult. People with albinism have vision that, in many ways, resembles that of a newborn baby. Even with glasses, their acuity, as measured using an eye chart, does not reach the normal 20/20 and may fall in the 20/40 to 20/200 range. A person with 20/40 vision sees at twenty feet what a person with 20/20 acuity can see at forty feet while a person with 20/200 acuity sees at twenty feet what a person with 20/20 vision can see as far away as two hundred feet. Given such poor sight, a person with 20/200 acuity is considered legally blind.

Anyone first studying the human eye may be surprised by the way its structures are arranged. The anatomy seems backward. Our retina, the light-sensing part of the eye, isn't located toward the front of the eye, where light enters, but is found in the back of the eyeball instead. What's more, the retina contains several layers of cells and neuronal processes, and the light-sensing cells, the rods and cones, are located in the layers almost furthest back. Rod cells are important for vision in low light and cone cells for color vision. These cells contain the light-absorbing compounds photopsin and rhodopsin, which are present in people with albinism. So, photons of light must pass through numerous structures and cells after entering the eye and retina before being absorbed by the light-sensing pigments in the rods and cones. This arrangement works because most of the eye's internal structures are transparent to light. However, as the baby grows, several changes occur in the retina that allow for high-acuity adult vision.

In the months after birth, the central part of the retina folds inward toward the back of the eye in the shape of a pit. This increases its surface area, allowing for the accumulation of more light-sensing cone cells. Indeed this change in shape has given the central retinal region its name, the fovea, which is the Latin word for "pit." Only cone cells are found in the fovea, although both rod and cone cells are found outside this region. As the retina matures, more and more cone cells migrate into the foveal region from the retinal periphery so that cone cells end up more tightly packed in the fovea than anywhere else in the retina. The outer segment of cones, the part of the cell that contains photopsin, also grows longer in foveal cones than elsewhere in the retina. What's more, all the other overlying retinal cells and processes move away from the front of the foveal pit. In this way, light hitting the retina at the fovea is transmitted directly to the cones without having to pass through other cell layers.[3]

It is not surprising then that we aim our eyes directly at an object to see it most clearly. When we look directly at a target, its image is cast on our foveas, which provide us with our sharpest vision. To appreciate the acuity of your foveal vision, try reading these words while looking straight ahead but holding the book a little to your right or left. In this way, you are looking at the letters, not with the fovea, but with more peripheral parts of your retina. You can see the letters; they are not exactly blurry, but they are not well resolved. You would have to increase the font size of the letters to read them.

But the fovea doesn't develop normally in people with albinism, like Liam, and, again, the absence of melanin is to blame. This pigment is found not only in the iris but in the retinal pigment epithelium, a sheet of tissue at the very back of the retina that envelops and nourishes the rods and cones. These epithelial cells are chock-full of melanin granules. As in other pigmented parts of the body, cells in the retinal pigment epithelium make their own melanin from the amino acid tyrosine in a series of chemical steps. Tyrosine is first converted into a compound called DOPA and then into several other molecules before its final conversion to melanin, and this pathway is deficient in many forms of albinism. As the retina develops, DOPA and melanin may be very important for the formation of the foveal pit and the migration of retinal cells.[4] Without these compounds, the foveal pit does not form normally, fewer cone cells move into the fovea, and fewer rod cells are found in other parts of the retina. So the retina in a person with albinism, with a shallow or nonexistent foveal pit and more loosely packed cones, resembles that of a newborn infant.[5] Without a well-formed fovea, visual acuity is compromised in a way that glasses cannot correct.

Liam used what vision he had. To inspect a new toy, he would bring it close to his face, so close that it was almost touching the corner of his right eye where he could see objects best. He would examine the toy thoroughly at this close range, going over every detail. Later, when he played with it, he did not try to see it but relied on his memory of it instead. Every time Cindy moved a toy to a new spot, Liam watched closely and then memorized exactly where that toy had been placed. I have yet to meet a person with significant vision problems or blindness who does not have an exceptional memory. Liam's memory was honed from an early age.[6]

Still, Cindy needed to know just how much Liam could see. One day, when Liam was two, Cindy quietly slipped into her bedroom.

Shortly afterward, Liam started looking for her, going methodically from room to room, guiding his movements more by memory and touch than vision. He would enter each room and call, "Mommy," but Cindy didn't answer. When Liam got to Cindy's bedroom, he came right up to her and asked, "Mommy?" Cindy still said nothing, and Liam turned away, continuing his search. When Liam finally came back to the bedroom and called again, Cindy answered. She has never forgotten this incident. It tortured her then and still pains her now to have not answered when Liam called, but she had to find out just how well he could see.

Between the ages of seventeen and thirty-four months, Liam saw the pediatric ophthalmologist four times. He hated the doctor. Imagine how painful it must have been for him during exams to have a bright light directed into his eyes. On the second or third visit, upon Cindy's insistence that the doctor address Liam's strabismus, the ophthalmologist patched one of Liam's eyes and left the room. Liam sat quietly in Cindy's lap, without reaching up to fuss with the patch. Then the doctor returned, and the sight of him so frightened Liam that he instantly burst into tears. The doctor ripped off the patch and told Cindy that, given Liam's poor vision, patching or any other treatment for strabismus would not work anyway. That was the end of that.

With strabismus, the two eyes are not looking at the same place in space so they provide conflicting input to the brain. A person with strabismus might adjust by squinting one eye shut. Some of my baby pictures show me dealing with my strabismus in this way. When my father's left eye crossed in his eighties, he'd look at me with his right eye wide open and his left eye almost shut. So, it's not surprising that in Britain strabismus is called by the unflattering name "squint." Liam too had to adjust to his misaligned eyes, which he did by habitually closing his right eye, thus looking almost exclusively with his left. To see anything out of the right eye, Liam had to close his left eye completely and then raise his right brow hard to move the right eyelid out of the way.

But why would a person with albinism develop strabismus, or misaligned eyes? While strabismus is found in about 4 percent of the general population, it is much more common in people with albinism. Strabismus may develop for a different reason in people with albinism because of a difference in the way the eyes are wired to the brain.

When you reach for a cup with your right hand, neurons on the left side of the brain fire to initiate and direct the movement. When your right hand contacts the cup, sensory signals are sent back to the left side of the brain. Tap your left foot, and neurons in your right brain are active. So, motor control and sensory processing for one side of the body occur in the opposite (contralateral) side of the brain. By analogy with your limbs, you might think that information from the right eye will be processed by the left brain and vice versa. That is indeed the case for animals, like rabbits, with laterally placed eyes. But it's not true for animals, like ourselves, whose eyes face forward.

Since our eyes are located on the front, not the sides of our face, each eye sees both the right and left halves of the visual field. Look out of either eye alone, and you'll see what I mean. So, it wouldn't make sense for input from the two eyes to be processed separately by opposite sides of the brain. Instead, visual information from the left side of the *visual field* is processed by the right half of the brain, and information from the right *visual field* is processed by the left. Imagine light rays traveling toward you, perhaps reflected by a bright object off to your left. These rays travel in a straight line, hitting the right, not the left, sides of both your forward-facing retinas. So, light emanating from the left visual field hits the right side of both retinas and is processed by the right side of the brain. The reverse is true for light coming from the right. As a result, input about a single object, as seen by both eyes, converges on the same neurons in the visual parts of the brain.

How, then, are our eyes wired to the brain to allow for this arrangement? Nerve fibers leave the retina in the optic nerve and pass through a brain region called the optic chiasm on their way to other

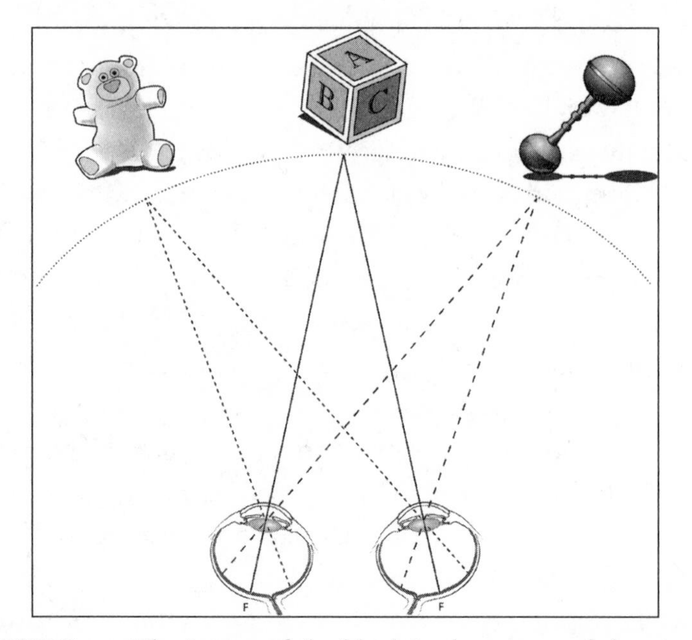

FIGURE 1.1. The image of the block in the center of the visual field is cast on the center of both retinas, while the image of the rattle in the right visual field is cast to the left of center and the image of the teddy bear in the left visual field is cast to the right of center on both retinas. *F* in the illustration stands for fovea.

visual areas. At the optic chiasm, there is a partial crossing over (partial decussation) of nerve fibers traveling from the retina to the rest of the brain.[7] Input coming from the right visual field excites the left retina of the left eye and is sent to the left side of the brain without crossing at the optic chiasm. But this same input also excites the left retina of the right eye and must cross over at the optic chiasm to the left brain to merge with the input from the left eye. The reverse is true for visual stimuli coming from the left visual field. So, in people with normal vision, about half the fibers coming from each eye cross to the other side of the brain at the optic chiasm.

However, the retinal nerve fibers don't follow these paths in a person with albinism. Some fibers that should stay on the same side of the brain cross over instead, and the extent of aberrant crossing over may vary from person to person.[8] This means that visual information

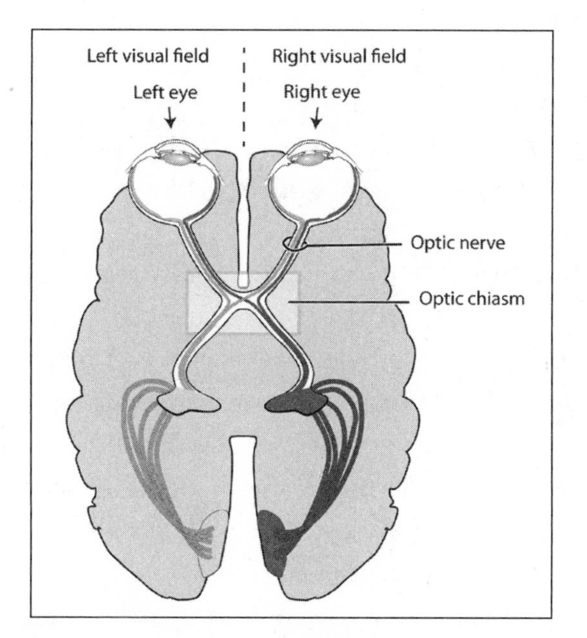

FIGURE 1.2. Information from the right side of the visual field is processed by the left side of the brain and vice versa. To accomplish this, approximately half the fibers coming from each retina cross over to the other side of the brain at the optic chiasm (indicated by the boxed area).

from part of the right visual field goes to both the left and right sides of the brain. The same is true for input from the left visual field. This misrouting is correlated with the lack of pigment in albinism and may lead, in turn, to disturbances in eye-movement control, causing misaligned eyes and an inability to see in 3D.[9] Testing, done when Liam was older, showed that he had excessive crossing over, and this may have contributed to his strabismus.

But we have to be careful in using our incomplete knowledge of vision and visual development to predict what someone can see. People with albinism don't have trouble, for example, seeing on which side of the visual field an object is located. Some people with albinism and the eye-brain misrouting have straight eyes and some degree of stereovision, and they are able to use their stereovision to judge depth and size.[10]

TWO YEARS AFTER LIAM WAS BORN, CINDY HAD A SECOND CHILD, a little boy who suffered from different but difficult medical complications. His pediatrician recommended that they see a specialist, not in their hometown of Columbia, Missouri, but at Children's Hospital in St. Louis. So Cindy asked if the doctor knew of a good pediatric ophthalmologist there. "As a matter of fact, I do," said the doctor, who then went off to his office to arrange appointments with both doctors on the same day.

It was a long, exhausting day in April when Liam and Cindy first met Dr. Tychsen. The morning was taken up with testing and consultations for Liam's brother. When they finally got to the ophthalmology offices in the afternoon, the receptionist took one look at Liam and asked about his glasses. Cindy told her that Liam didn't have glasses, and the receptionist looked surprised, commenting that children like Liam often have glasses by his age. Even the receptionist, Cindy thought, knew that Liam should have had treatment by this time. Then Liam went through a host of vision tests before they met Dr. Tychsen. Through it all, Cindy worried what the backlash from this long afternoon would be. After each of their four visits to the previous ophthalmologist, Liam had withdrawn completely, not speaking to anyone but his mother for two weeks. Yet the contrast between the two doctors could not have been greater. Dr. Tychsen was a quiet presence. He never rushed them, and he never insisted. As they left his office, Liam asked, "Mommy, can we come back tomorrow and see Dr. T, the doctor who loves me?"

Dr. Tychsen explained that Liam was living in "a cocoon of visual blur." His zone of clear vision extended out only three inches from his nose. Liam's visual impairments were a mixture of three disorders: extreme nearsightedness (pathological myopia), strabismus (eye misalignment causing double vision, lack of depth perception, and visual confusion), and albinism. The myopia was treated by prescribing thick glasses, which improved Liam's distance vision from 20/2000 to 20/200. Dr Tychsen addressed Liam's strabismus by performing a sequence of eye-muscle surgeries at ages three, five, and seven. After the first surgery and, indeed, after every intervention in Liam's vision, Cindy, a most wise, observant, and devoted mother, was repeatedly taken aback by Liam's reactions. These were often revealed by a casual remark that Liam would drop here and there. After the first strabismic surgery, Liam asked, "What did you do with my other mommy—the fun one who stood behind you?" Before surgery, Liam had double vision. He saw the second mommy image above the first, and it appeared to walk on tables and play in the air.

Liam was not entirely blind, but his visual development had been severely disrupted. Even with his glasses, he couldn't see faces or objects a few feet away or understand spatial layout. Thirteen years after Liam's first visit, Dr. Tychsen, through a new series of operations, was able to give Liam near-normal eyesight. Yet Liam's lack of visual experience throughout childhood had a long-lasting impact. As we shall see, the operations were only the beginning of Liam's vision restoration.

As a child, Liam couldn't distinguish a cat from a dog. Each was, as Liam later wrote to me, "a living thing that was on the ground and had hair." When Liam looked at someone's face, the mouth and nose blended together into one big blur. Eyes were two dark spots. So Liam didn't learn to recognize faces or facial expressions. He identified people by their hairline, skin color, and clothes, and this placed severe restrictions on what Cindy could wear. One morning Cindy shed her usual dark blouse and pants and dressed up in a skirt and boots to give a speech at church. When she came down to the kitchen, Liam got very upset and demanded to know what had happened to his mother. Cindy assured him that everything would be all right and kept the nice clothes on, just this one time, for church. During one of their trips to the medical center in Columbia, Liam, mistaking a strange lady in dark clothes for his mother, followed her into an elevator. After a few panicked moments, Cindy found Liam with a security officer in the elevator lobby on another floor. From that day onward, Liam always grabbed his mother's and his brother's hands before entering an elevator, insisting, as the protective older sibling, that they must all hold hands lest his brother get separated from them. Given Liam's limited vision, Cindy knew better than to ever try to wear a hat.

But poor vision did not prevent Liam from learning to ride a bike. When Liam was five and a half, a kid-sized bike showed up for Christmas. It was a mystery gift, and, to this day, neither Cindy nor Liam knows who gave it to him. The shiny new bike was big for Liam, but he learned to ride it anyway in a nearby parking lot, even sliding down the December snow banks the way other kids did.

One day at about this time, Cindy took Liam out of the car, put him down at its back, and walked toward the front. Liam froze. He could no longer see Cindy. His bubble of vision allowed him to see clear images only very close up and to make out blurry images no further than four feet away. Cindy wondered what would happen when Liam grew to be more than four feet tall and could no longer see the ground from his bike. Sure enough, he stopped riding.

As the school years approached, Liam became so shy around people outside his family and a couple of friends that he would do everything asked of him but speak. His kindergarten teacher was thrilled when she finally got Liam to respond to her questions with a simple "yes" or "okay." In conversation, we are always observing other people and adapting our movements to theirs. We copy each other's facial expressions, eye gaze, and body movements. Liam couldn't see these actions, making social interactions more difficult and painful.

Cindy insisted that Liam be provided with Braille instruction, so in kindergarten Liam was also given a teacher for Braille and "O and M" (orientation and mobility training for the blind, including the use of a white cane). It took about a year before Liam warmed up to the teacher and could start learning Braille. In fourth grade, Liam's classroom teacher turned out to be his familiar O and M teacher. At this point, Liam decided, begrudgingly, to speak in class, not because he wanted to but because it was easier to give in and talk than to remain silent. But Liam still hated writing essays for the teachers. They seemed too personal.

Although Liam learned to read both Braille and the printed word, he concentrated on printed text. In first grade, with his glasses, his acuity was good enough to read

20-point font.

Since most books for young children are written in a font that size, Liam enjoyed reading. Had albinism been the sole cause of Liam's vision problems, however, his acuity should have remained stable. But other factors must have been involved because Liam's myopia (nearsightedness) continued to get worse. So, as he got older, he needed a larger font size to read, but the print in schoolbooks got smaller.

If you have myopia or hyperopia (farsightedness), then light rays focus not on the retina but in front of or behind it, respectively. To see most clearly, you are given prescription lenses that focus or bend the light onto the retina, and the optical power of the lens is measured in diopters (D). A prescription of –1.0 D will correct a mild myopia, while a prescription of +1.0 D will correct a mild hyperopia. Most infants start out with a mild hyperopia of approximately +1.0 D that decreases during childhood. If you know people who are very nearsighted, the ones who got glasses in the fourth grade, they probably have a prescription of around –6 to –8 D. But at age two, Liam required a much stronger lens, –14 D, which increased further to –18 D by age nine. At age twelve, his prescription was –20 D, so that even with glasses, in his middle school years, Liam required

70-point font

to read. He had to reprint assignments in large font on several pieces of paper, tape them together, and then fold the whole structure up four times to make it the size of a normal sheet.

Liam's glasses were a sight in themselves. His lenses were so thick that the arms of the glasses could not fold flat, which meant that the glasses didn't fit into a standard glasses case. Traditional earpieces weren't strong enough to hold up the lenses; Liam used wraparound cables instead. The edges of the lenses were so sharp that they would cut Liam's cheeks, and the lenses were so thick that they created optical distortions. Yet the double concave lenses were actually one-third to one-half as thick as they would have been if they had been made from standard materials. They were made instead with a special, high-refractive-index plastic in a lab in Kansas City. "They were tough," Liam told me. "If they were dropped, they didn't get hurt, they hurt whatever they landed on."

During this time, Liam didn't so much read as decode. He saw letters as round or square and divided them further into seven shape categories. Small *c* and small *e*, for example, were in the same category since they had a similar shape. Then, he would guess at a given letter and keep it in memory as he decoded the next. The letter *t* sticks up at the end so a three-letter word with a letter that stuck up at the end could be the word "cat." Obviously, context played a huge role in deciphering words. Reading was tedious, and Liam started to hate it.

Not surprisingly, homework took forever. Liam began it on the bus ride home by listening to audiotapes. Like many people who are blind or have low vision, his hearing was more sensitive and his auditory processing faster than normal. He found that most people turned up the volume too loud. When listening with a screen reader, a tool that translates the words on a computer screen into speech, he would adjust the word rate to a speed most of us couldn't follow. Once home, he'd work until 11 p.m. and then get up early to finish his assignments.

Fortunately, he had a phenomenal memory. When he was a child, his mom would read him a page from a book, and then Liam, who hated sitting still, would jump off the sofa and repeat the entire page verbatim. In middle school, he was in honors math despite the fact that he couldn't see the decimal point in numbers, even when they were blown up to a large font. Liam could remember and work out long operations of numbers in his head. Like others with severe vision loss, he developed an exceptional working memory to help compensate for a lack of sight.

But all this effort took its toll. As the school day wore on, Liam's vision would degrade, and colors would fade. Red was the hardest color to see. It would fade to brown, while blue, Liam's favorite color, faded least. Even when his vision was at its best, he had a hard time telling orange from red. Liam's problems with the color red and his preference for blue may derive from the way color is sensed by our visual system.[11] We see reds and greens most clearly in the center of our vision. Indeed, in the very center, the area seen by the fovea, we see reds and greens best and do not see blues at all. On the other hand, blues are seen more evenly across the whole visual field, with the exception of the very central area. When the reds and not the blues began to fade as the school day wore on, Liam may have been losing more and more of his central vision.

Liam's math teacher noticed his color struggles one day when they were working with colored blocks. He asked Liam what block could be removed from a pile without changing the appearance of the pile from the front or side view. Liam could solve the problem, but he couldn't distinguish the blocks by color.

But only the math teacher seemed sensitive to Liam's struggles. He was stuck between two worlds—not completely blind but hardly able to see. Being blind would have made it much easier because then the school would have provided him with Braille textbooks. The reading specialist for the middle school had been taught that people with albinism had reduced but stable visual acuity, so to read, they just needed larger print. She disregarded Liam's severe myopia and dismissed his need for accommodations, claiming that he was just acting out. The school dictated how and what Liam should learn, and he had to adjust to those demands.

Cindy, ever supportive and resourceful, tried to mitigate these struggles with a special family tradition called "dark night." Once a week, they would turn off all the lights and then eat dinner and play Braille Monopoly in the dark. Liam, Cindy reasoned, was almost always at a disadvantage except on "dark night." By relying less on vision and more on tactile and auditory cues and spatial memory, he could move about better than anyone else.

"How far is your vision?" Cindy remembers Liam asking her. "'Twas a physical sensation of being knocked in the chest," Cindy wrote when she realized the import of Liam's question. "I can actually still visualize the color of the sky that day and what I saw as I looked around trying to capture a limit on my vision." There was no limit. At night, she could see stars that were lightyears away. For Liam, distant things didn't just get blurry. He didn't see them at all. Good vision allows us to perceive great vistas and locate distant objects. Liam had no concept of this.

When Liam was twelve, with his vision degrading, Dr. Tychsen suggested a lensectomy (or clear lens extraction), an operation that would have removed the natural crystalline lenses from Liam's eyes. Light is focused or bent by two structures in our eyes, the cornea and the lens. Liam's eyeballs were so long that light reflected off objects and into his eyes focused far in front of his retina. With removal of the natural lens, the light would be bent only by the cornea and would focus further back, at the distance of Liam's retina. This would improve Liam's acuity and, as Dr. Tychsen pointed out, would eliminate the need for thick glasses to see further distances. But without his native lenses, Liam would no longer be able to focus at near distances, requiring the use of bifocals or reading glasses. Liam and Cindy considered the operation but, in the end, rejected the idea.

One day while Liam was in middle school, he called his mom, complaining that he was cold and needed a sweater. Later in the day, they went to the horse stables to watch Liam's brother ride, but Cindy noticed that Liam was feverish. So they went straight home. Liam's fever was higher than 105 degrees, and Cindy struggled to bring it down.

Shortly before this fever, Liam had met with the reading specialist for tests. All was OK. Three weeks later, there were more tests, but Liam had been through the fever during that time. During the second round of testing, he couldn't read at all. He had no comprehension and no memory. In eighth grade, Liam had been in honors math, but in ninth and tenth grades, he struggled. His speech was reduced to sharp, short sentences. Formerly athletic and continually in motion, Liam felt tired and weak all the time. He had always been slight but now got heavy. He would cry. This was astonishing to Cindy because Liam never cried. Even though his hair was falling out, many people told him that his problems were all in his head, and the reading specialist still maintained that Liam was acting out and needed no further accommodations.

Despite trips to a neurologist and other specialists, it took four long years to diagnose what was going on. Finally, an endocrinologist figured things out. Liam had hypothyroidism. The high fever he had suffered, perhaps produced by a virus, had destroyed his thyroid. He started taking the thyroid hormone thyroxin and began to improve. Indeed, he started piano lessons, progressing so rapidly that he learned in one year what is normally mastered in six. At one piano recital, he ran into his endocrinologist. When the doctor first treated Liam and heard his short, spare speech, he had thought that Liam must be developmentally delayed. "If you can play the piano like that," the doctor now asked, "what else can you do?"

Liam's perceptual world was shaped by sound, touch, and spatial memory. With these skills, he compensated for his vision loss so well, it was not just the school reading specialist who underestimated what he could see. Liam knew who was coming to the house before they arrived because he could recognize the sound of each person's car motor long before most of us could hear anything at all. He moved through his house in the light or dark with ease. The same was true at church until the church added a new addition, and Liam crashed into a glass door. Even Liam's grandmother did not realize how little he could see.

Yet, as high school began, Liam used his vision less and less. His refractive error had increased to –23.5 D, so that, even with glasses, he was seeing only 20/250, worse than legal blindness. To prevent his tripping over things, he shuffled his feet when he walked. When he reached for a glass across the table, he would slide his hand along the surface until he felt the glass. It took him so long to focus on anything that he could no longer track moving objects. One day when Cindy went to pick Liam up from school, she watched him talking with a group of classmates. While the other students looked at each other as they chatted, Liam stood with his head down, holding his white cane. He's virtually blind, Cindy thought, and this condition, it seemed at the time, would follow him for the rest of his life.

– chapter 2 –

Dr. Ridley's Brainchild

"WE MAY HAVE ANOTHER CANDIDATE," CINDY OVERHEARD the doctors whisper during one of Liam's vision exams in 2004. Later that same day, Cindy and Liam were told about a new type of surgery involving intraocular lenses (IOLs). This new procedure would add a second lens to Liam's eyes. The artificial lens would help Liam to see more clearly at a distance, but he would also keep his natural lens, allowing him to focus on objects both far and near. When the two returned home, Cindy spent hours on the internet investigating the procedure. At a subsequent appointment, Dr. James Hoekel, an optometrist who worked closely with Dr. Tychsen, explained the surgery in great detail and sent Cindy and Liam home with his personal cell phone number, encouraging them to call with any questions they might have.

Liam, now fifteen, was old enough to be part of the decision. Still fighting fatigue from an undiagnosed thyroid deficiency, he was hesitant. What's more, he liked his glasses. Sure, they were thick and heavy and held on his face by a wraparound cable, but, Liam told me, he liked extreme things. He and his mom didn't regard his visual impairment as an awful, consuming condition to be mourned and regretted and that needed to be "fixed." Liam's compromised vision was an essential part of him. It was just a trait to work around, stimulating them to find different ways to accomplish things. They were not looking for a miracle cure. But the intraocular lenses would improve Liam's visual acuity, thus providing some practical benefits and ease of function. So, ultimately, they decided to go ahead with the operations.

An unlikely sequence of events, involving World War II planes and pilots, led to the first intraocular lenses, and those lenses led, decades later, to the current IOLs that radically changed Liam's sight.[1] On August 14, 1940, during Hitler's attack on Britain's Royal Air Force, flight lieutenant and flying ace Gordon "Mouse" Cleaver was flying back to his base when a bullet smashed through the cockpit of his plane. He had been in such a rush earlier that day that he hadn't put on his flight goggles, a tragic mistake because flying fragments from the cockpit canopy immediately blinded him in both eyes. Remarkably, Cleaver managed to turn his plane upside down and parachute to safety. He was picked up and rushed to an infirmary for treatment.

Cleaver received eighteen surgeries on his face and eyes, and most, if not all, of his eye surgeries were performed by ophthalmologist Harold Ridley. One eye was lost, but the other could be saved, even though several fragments of acrylic plastic still remained scattered throughout it. Ridley followed Cleaver for eight years and noted that his eye did not react to the plastic. Indeed, in other pilots too, plastic fragments from smashed cockpit canopies did not induce reactions in the eye. The material appeared inert.

After the war, Ridley returned to more routine eye surgeries such as cataract removal. But even after his patients recovered from surgery and were outfitted with thick, coke-bottle spectacles to make up for the removed lens, their eyesight remained poor. In 1948, one of Ridley's medical students, Stephen Perry, asked him if he was going to put a new lens in a patient's eye after the cloudy, cataractous lens was removed. This idea had been in Ridley's mind for some time. He thought back to the pilots and the plastic in their eyes. Could a lens be made from this same material and then inserted into the eye?

Ridley contacted Rayner and Keeler, an optical company, and had the first intraocular lens made using the same acrylic plastic that was used in airplane canopies. On November 29, 1949, he operated on Elisabeth Atwood, a forty-five-year-old hospital nurse who had a cataract in one eye. He inserted the artificial lens but removed it shortly afterward. He then reinserted it into Atwood's eye on February 8, 1950. This was the first time an artificial lens had been used to replace the natural lens, yet Ridley did not photograph or film the operations or mention the plastic lens in his surgical notes. He told Atwood to keep the lens a secret, fearing great objections from his colleagues.

Dr. Ridley's fears were totally justified. In July 1951, after having operated successfully on several more patients, he announced his findings at the Oxford Ophthalmological Congress. His talk was met with little interest and even with hostility from many of his colleagues. Ophthalmologists at the time were used to removing foreign bodies from their patients' eyes, not inserting new ones! Indeed, it took another thirty years before intraocular lenses, many made with acrylic plastic, were routinely used to replace the natural lens in cataract operations. Finally, at age ninety-two, after a lifetime of frustration, Harold Ridley was knighted by Queen Elizabeth II. Not only did his artificial lens improve the sight of millions of cataract sufferers, but his innovation spurred the invention of many other prosthetic devices, such as artificial heart valves and limb joints.

Ridley used his artificial lens to replace a natural lens that was clouded with cataracts. Could a similar lens be used in people with severe refractive error (nearsightedness or farsightedness)? Such a lens would need to compensate for changes in the length of the eyeball that result in poor acuity. This is what normal eyeglasses also do, but if eyesight is really poor, the spectacles become so thick that they introduce other optical problems. An artificial lens introduced directly into the eye could compensate much more effectively for such severe acuity loss. But one problem remains. Seeing well requires a lens that can focus at many different viewing distances. When we look from far to near, our natural lens changes shape. Ridley's artificial lenses, on the other hand, could focus only at one distance.

In 1989, eye surgeons in Europe began to insert intraocular lenses into the eyes of patients who had severe refractive errors rather than cataracts.[2] The surgeons did not remove the natural lens, however; they inserted a second lens into the eye. This allowed the patient to use his or her natural lens, with its ability to change shape, along with the artificial lens to see clearly at all viewing distances. An improved version of this artificial lens was first approved for use in the United States in 2004. Here's where Liam and his surgeon, Dr. Lawrence Tychsen, come back into the story.

When Dr. Tychsen was a young physician, he examined a child with severe cerebral palsy who was considered blind. Dr. Tychsen discovered, however, that the boy's visual system was actually working and told a more senior doctor that this child needed a good pair of glasses. The older doctor responded callously, "Potatoes have eyes, but they don't need glasses." Although shocked by this cynicism, Dr. Tychsen knew then that he wanted to work, was even called to work, with patients labeled by others as too disabled or with problems too difficult to help. Today, his practice includes many young children with severe neurological and visual disorders, and Dr. T. doesn't give up on any of them. Some of his patients have such poor acuity that they live in what he calls "a cocoon of blur where visual stimuli are noxious and frightening."[3] Starting in 2005, Dr. Tychsen implanted intraocular lenses in his young patients who couldn't benefit from or tolerate eyeglasses or contact lenses and whose vision couldn't be corrected with laser surgery.[4] Liam was one of these patients.

– chapter 3 –

A Window on the Brain

LIAM UNDERWENT SURGERY ON HIS FIRST EYE IN DECEMBER 2005, when he was fifteen years old. Surgery on the second eye followed five weeks later. There was no sudden "I can see!" moment. Given his hypothyroidism, a condition that slows body metabolism, it took Liam longer than usual to recover from the anesthesia. His eyesight was supposed to clear up in six weeks. Instead, it took months. But when it did, Liam's visual acuity had vastly improved. Before surgery his vision was 20/2000 without glasses and 20/250 with the thickest lenses. Six months later, with no glasses, he was seeing 20/50. His albinism prevented him from seeing 20/20.

Despite his slow recovery, Liam's visual behavior changed within an hour of the outpatient surgery. With the nurse's encouragement, Liam tried to stand up but immediately fell over. He adjusted quickly, and once he was more stable on his feet, he and Cindy took a walk down the hall. A girl waved at them, and Liam asked his mom why the girl stuck up her arm and moved it around. Cindy was floored. When Liam was little, and the two were out walking, they routinely passed a bus driver who waved at them. Cindy knew Liam couldn't see the driver, so she would say, "The driver is waving at you. Wave back."

When Liam did, Cindy assumed that he understood the meaning of the gesture. But he had not. Liam didn't know what his action, his wave, looked like to other people.

Nine months after the second surgery, one of the lenses moved out of position, causing Liam to experience double vision. The lens was replaced, and this time the improvement in acuity was immediate. The gains in acuity after the first two surgeries may have occurred gradually because it took the brain some time to process all the new information the eyes could now provide.

The improvements in Liam's vision were far more dramatic than his doctors had predicted. Not only did his acuity improve tremendously, but his color vision became normal. No longer did reds fade for Liam as the day wore on (though his favorite color remained blue). Nystagmus, or an involuntary back-and-forth movement of the eyes that is common in albinism, was reduced. His binocular vision improved, as did his depth perception, albeit slowly.

But the improvements were discombobulating. Surgery plunged Liam into a world of sharp lines and edges. He missed the soft color blends that he used to see. After the surgeries, with his vastly improved acuity, Liam saw lines particularly where there were changes in color, lightness, or texture. This could occur within or between objects. He saw lines where one object ended and another began, where an object in front occluded an object behind, or where a shadow was cast on a surface. While we all see lines at object or shadow boundaries, we know where these lines belong. But after a childhood of near-blindness, Liam did not recognize the lines as boundaries of known objects. Instead, he saw a tangled, fragmented world.

As Liam and his doctors were to learn, the ability to see lines and colors clearly is not enough to make sense of the visual world. With each glance, we don't usually experience isolated features like lines and colors because they arrange themselves into people, animals, objects, and landscapes. We recognize an object immediately—all its parts combine together, instantly and effortlessly, into a single unit. Indeed, experiments demonstrate that we can determine whether or not a picture shows an animal even if we have seen the picture for only one-fiftieth of a second.[1]

Yet, for Liam, to construct a scene from the lines and colors he saw required constant attention and analysis. While his ability to recognize objects has improved over the years, this is how he described his first views after his operations:

> I think of a line as the difference between two colors, the point where two colors meet. These lines make up everything I see . . . A surface is pretty much consistent until you come across a line. If there is a line on a surface it could mean a change in horizontal to vertical like the corner of a room, or a change in depth like the drop-off of a stair or curb, or it could also mean nothing important pertaining to the physical structure, like a crack between floor tiles or sidewalk squares . . . In addition to all of this there is also some information that is completely useless that other people can filter out. I probably cannot fully explain it in terms you are used to as a sighted person, but there are things that light can do that can add lines to any surface or multiple surfaces (which I affectionately call "lies"), and I have to figure out what lines to disregard and which are pertinent in addition to deciphering what they mean.

Liam's dependence on color to mark out the lines and sort out what he saw is common among people who gain sight in adulthood.[2] Seeing that two objects are different colors does not require prior visual experience. Following a corneal stem cell transplant, Michael May saw for the first time after forty-three years of blindness, and his first visual experiences were of the colors of his wife's and doctor's clothes and face.[3] Sheila Hocken, blinded when young by cataracts, was astonished by the colors she saw when her surgical bandages were first removed. These colors, the colors of the uniforms worn by the hospital staff, looked brilliant to her.[4] SB, the man studied by Richard Gregory and Jean Wallace who gained sight at age fifty-two, was also initially stimulated by colors, as was Virgil, the sight-recovery patient whom Oliver Sacks described.[5] Using color to isolate and identify objects has its pitfalls, however, since many objects are made up of more than one color.

It was hard at first for Liam to recognize a change in depth, so for something as routine as walking down a sidewalk, Liam had to use his sight with caution. If he saw a line in the sidewalk, he had to judge whether it indicated the junction between flat sidewalk blocks, a crack in the cement, the outline of a stick, a shadow cast by an upright pole, or the presence of a sidewalk step. Should he step up, down, or over the line, or should he ignore it entirely? "So while I walk," Liam wrote, "I am always focusing on the lines I am seeing and calculating what moving across the next line means and where I can step and what I shouldn't step on." Since he had to focus so intensely on the lines right in front of him, he couldn't develop a good sense of his overall surroundings, and this loss of context, as Figure 3.1 suggests, made his views that much harder to interpret. The amount of analysis that Liam employed on a glance-by-glance basis was overwhelming and exhausting.

FIGURE 3.1. While the edges of stones and bricks are clearly outlined in this photo, the flattened view, without a sense of the overall context, makes the scene more difficult to interpret. This photo was taken by the author while standing on a flagstone landing (bottom of photo) looking down onto the brick steps and flagstone path beyond.

Liam's postsurgical views contrast sharply with those of Robert Hine, who gradually lost his vision in adulthood but then recovered it after fifteen blind years.[6] Immediately after his sight-restoration surgery, he recognized his wife's face, the bottles and instruments in the examination room, and, upon returning home, the flowers in his garden. Thanks to his visual experiences in childhood, Hine, immediately after surgery, saw whole objects, not just lines and colors. Like most of us, he automatically integrated features into meaningful wholes—a face, a flower, a car.

Many of us have visited an art museum where every painting, every object deserves careful study. After an hour at the museum, we feel drained. The museum floor seems hard, the benches uncomfortable; yet there is so much more to see. This phenomenon even has a name: "museum fatigue."[7] To Liam, the whole world was like an art museum, where every piece required attention and analysis, demanding of him enormous stamina.

Liam's childhood vision may have been very poor, but it did provide him with some visual skills. People blinded from birth or infancy, who see for the first time, may not be able to recognize a simple shape. To tell a triangle from a square, they may need to count the number of corners; they do not see these shapes as a whole.[8] While many sight-recovery patients learn quite quickly how to recognize individual letters with their eyes, they need much more time and practice to see how the letters come together to form words. This was true even if they had been adept at reading Braille. So, for shapes and words, it was difficult for the newly sighted to see the whole from the arrangement of its parts.

Liam could recognize simple, flat geometric shapes drawn on paper and could read print. Yet these skills, developed in childhood, didn't translate into recognizing large objects, like chairs, arranged in three-dimensional space. Indeed, Liam's description of lines and colors resembles the first views described by a congenitally blind woman who gained sight at age twenty-five: "all around me I see an ensemble of light and shadow, lines of different length, round and square things, generally like a mosaic of changeable sensations that astonish me, and whose meaning I do not understand."[9]

The more often we see and identify an object, the easier it is to remember it. Liam too wondered if he could recall the look of an object from memory. One morning in 2011, six years after his surgeries, he tried his hand at drawing by tracing a few simple cartoons. Later in the day, he realized he could use these drawings as a test of his visual recall. Could he reproduce them from memory? He could indeed, as shown in Figure 3.2, but it was his description of his mistakes that surprised me.

FIGURE 3.2. Liam's sketch of a flower, drawn from memory, is shown to the right of the original cartoon.

Liam's drawing of the flower on the right looks similar to the original on the left. He writes, "I originally had the stick coming from the center of a circle and changed it because I was sure it was wrong and it came out between circles." The circles represented the flower's petals, but he described these structures not as parts of a real flower but as geometrical shapes.

FIGURE 3.3. Liam's sketch of a cat, drawn from memory, is shown to the right of the original cartoon.

Similarly, in his drawing of a cat, Liam wrote that the triangle was missing (Figure 3.3). The triangle represented the cat's nose, but, again, he identified it as a geometric shape, not as a part of the face.

These remarks suggest that, even six years after his surgeries, Liam often saw objects not as wholes but rather as combinations of lines and geometric shapes.

Indeed, at this time Liam was in college, and he talked a great deal with his computer science professor, who saw similarities between Liam's recently acquired vision and computer vision. The professor noted Liam's literalness and focus on details. One day, in 2011, Liam was in the professor's office when he noticed a bright shape on the office door. He asked his teacher why the door had a "circle with a cross on it," but the professor did not initially know what Liam meant. After some time, the professor realized that Liam was seeing a reflection on the glass door of a piece of paper that was lying on the floor. The professor saw the door as a whole. He ignored the extraneous light pattern caused by the reflection because it was not part of the door itself.

I first met and began my correspondence with Liam in 2010, five years after his surgeries, and visited him again in 2012 and 2014 when he was twenty-two and twenty-four years old. I was to learn that many of the skills we take for granted, such as recognizing objects or negotiating a flight of stairs, were visual puzzles for him. Always practical and analytical, Liam developed strategies to solve these puzzles. Over time, his vision became more automatic and less effortful, but seeing can still be a challenge for him, especially in unfamiliar places.

Liam's focus on details is very common among sight-recovery subjects and may help explain some of their responses to popular visual illusions.[10] In 2014, while visiting Liam, I brought with me several optical illusions, and together we used them to explore his vision. These illusions are considered "context-based" since, to be fooled by them, we need to see the whole picture. For example, in the Müller-Lyer illusion, the two lines look to be different lengths even though they are actually the same size (Figure 3.4). In the upper figure, the arrowheads point toward the line, while in the lower figure, they point away from the line. Most people see the upper line as longer. When Liam looked at this illusion, he saw the two lines as about the same size. So the illusion had no or only a weak effect on him. The same was true for SB, as reported by Gregory and Wallace, and KP, a man who was blinded at age seventeen and regained sight in one eye fifty-three years later.[11] In contrast, the children from Project Prakash and LG, a sight-recovery patient described by Valvo, were fooled by the illusion.[12]

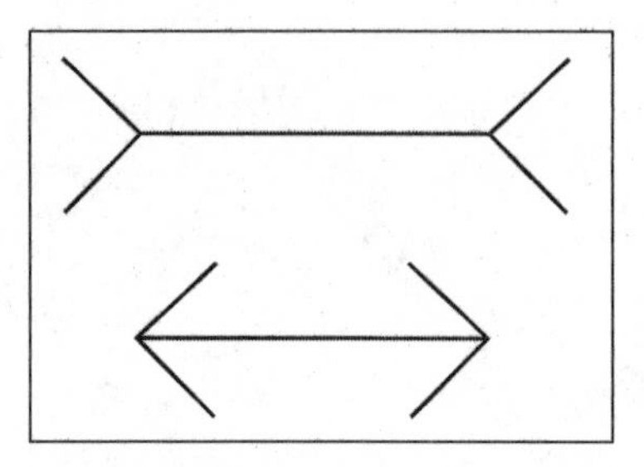

FIGURE 3.4. The Müller-Lyer illusion. Are the lines the same length?

Nor was Liam fooled by the Hering illusion (Figure 3.5). The heavy, vertical lines in this illusion appear curved. Take away the thinner, radiating lines in the background, and the heavy, vertical lines reveal themselves as straight. Liam saw the vertical lines as straight, as if he did not see them as part of the larger picture. SB was not fooled by this illusion either.[13]

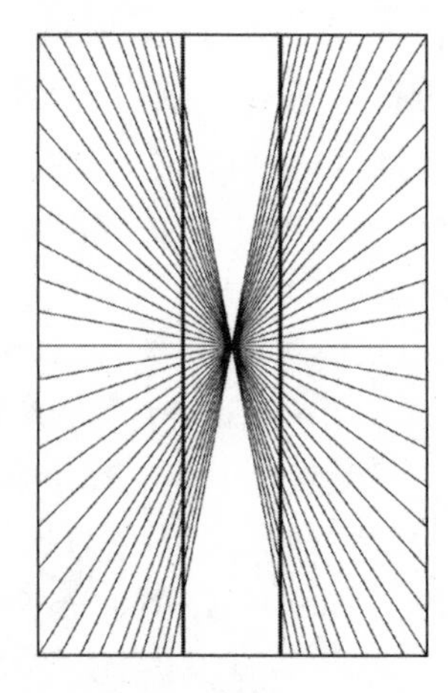

FIGURE 3.5. The Hering illusion. Are the vertical lines curved or straight?

Before his surgery, Liam told me, it was hard to tell a dog from a cat. They were just living, moving things that were on the ground and had hair. Now, I showed Liam a photograph of a dog together with a cat, and he was able to distinguish the two. However, he couldn't distinguish the dog from the cat in a picture of their silhouettes. Infants, at three to four months of age, can discriminate between silhouettes of the two animals, but, for Liam, the silhouettes may not have had enough individual features to differentiate them.[14] While he can see the parts, he still struggles to see the whole.

In 2015, one year after my third visit, I emailed Liam the picture in Figure 3.6 from Josef Albers's book *Interaction of Color*.[15] This picture illustrates how we see transparency. Most of us would interpret the picture as showing two sheets of paper, with the lighter sheet on the right being translucent and lying over and partially blocking our view of the darker sheet on the left. Liam's response was very literal: "Two rectangles . . . with an arrow shape in the middle of the two." He saw the picture as a flat abstract design of geometric shapes, not as representing two sheets of paper at slightly different depths. He had no trouble seeing the elements of the picture but did not grasp the meaning of the whole composition.

FIGURE 3.6. We interpret this figure to include two sheets of paper with one semitransparent and partially overlapping the other.

Although Liam can recognize many more objects today than he could right after his surgeries, he still struggles to see whole objects rather than their parts. While we give little thought to our ability to instantly segregate and recognize objects, vast numbers of brain neurons and networks are involved. Indeed, about one-third of our brain is devoted to vision and visual processing. Having been practically blind since infancy, Liam's visual system did not develop normally. His initial views, upon receiving his intraocular lenses, and his subsequent struggles give us some insight into what all those visual networks are doing and how much we take our vision for granted.

LIAM'S EMPHASIS ON LINES, EDGES, AND CONTOURS, RATHER THAN whole objects, makes a great deal of sense to a visual physiologist, for our visual system is indeed sensitive to lines. A line often marks the border between light and dark, and many output neurons of the retina respond best to this contrast.[16] The primary visual cortex, also called V1 or the striate cortex, is the first area of the cerebral cortex to receive visual input from the retina. Starting in the mid-1900s, David Hubel and Torsten Wiesel began recording the firing activity of neurons in V1, first in cats and then in monkeys.[17] Each neuron had a classic receptive field, meaning that it was sensitive to a pattern of light stimuli in a particular small region of the visual field. One neuron might be sensitive to stimuli located straight ahead at the center of the visual field, while another might respond, say, to light stimuli located a little to the left of and below that center point. Adjacent V1 neurons have receptive fields that are slightly different but overlap since they receive input from slightly different but overlapping regions of the retina. In this way, the whole visual field is mapped in a topographic or retinotopic manner onto the primary visual cortex. While some V1 neurons respond selectively to light of different wavelengths (what we perceive as colors), most V1 neurons respond to light bars against a dark background or to dark bars against a light background. The orientation of these bars matters. Some V1 neurons are most excited by vertical bars, others by horizontal bars, and still others by bars of intermediate angles. These are the orientation-selective cells.

When you see the edge of an object, you may see a change in lightness or texture right at that edge, and that contrast stimulates a subset of your V1 orientation-selective neurons. As you follow, with your eyes, the changing direction of the edge, different orientation-selective neurons in V1 respond. Long-range connections between V1 neurons with similar orientation preferences help bring out lines or contours spanning long lengths.[18]

Just like Liam couldn't make sense of the world using only features like lines and colors, we cannot understand visual processing by studying individual V1 neurons alone. In 1973, the great neuropsychologist A. R. Luria published a book titled *The Working Brain*.[19] Basing his conclusions on careful observations of his patients, Luria described our visual brain as organized into primary and secondary zones. The primary zone includes the primary visual cortex, or V1, and damage to a part of V1 results in blindness to the part of the visual field from which that V1 area receives input. Neurons from the primary visual zone communicate with neurons in the secondary zones, where the raw input is analyzed, combined, and synthesized. People with damage to their secondary visual zones are not blind but suffer from different forms of visual agnosia ("not knowing"). They may "see" all the parts of an object, its contours and colors, but cannot put the features together to recognize the object as a whole.[20] It's as if they have never seen the object before.[21]

Perhaps the most famous patient with visual agnosia was Oliver Sacks's patient Dr. P., "the man who mistook his wife for a hat."[22] When Dr. Sacks showed his patient a glove and asked him to identify it, Dr. P. responded, "A continuous surface infolded on itself. It appears to have five outpouchings." Dr. P. gave a fine description of the key features of a glove but couldn't put these features together to perceive the whole. It is not that he lost the concept of a glove, because he could recognize it through touch. His troubles and those of others with visual agnosia mirror Liam's predicament. Like Liam, they don't automatically and unconsciously recognize objects but must reason out their identity from basic features. In all cases, the problem involves the higher visual pathways in the brain.

V1, with its orientation-selective cells, is located along the visual pathway between the retina in the visual periphery and the visual areas located higher or more centrally in the brain.[23] In the peripheral (toward the eye) direction, V1 neurons receive their input from neurons in the thalamus, which connect directly with retinal cells. In the central (further into the brain) direction, V1 neurons communicate with cells in V2 (visual area 2) and then directly or indirectly with neurons along two major visual pathways. The "what" or "perception" or ventral pathway is involved with the recognition of objects, faces, and places, while the "where" or "action" or dorsal pathway is concerned with object localization and guidance of movement.[24] Luria's "secondary zones" correspond to these pathways. While V1 and V2 are part of both the perception and action pathways, higher regions may contribute more to one pathway than the other. Higher areas are labeled in different ways. Some are named with *V* followed by a number that suggests their place in the visual hierarchy. Others are named for their anatomical location. Area V4 and the lateral occipital region, for example, are part of the "perception" or "what" pathway, while V5, also called MT for medial temporal region, is part of the "action" or "where" pathway (see Figure 3.7).

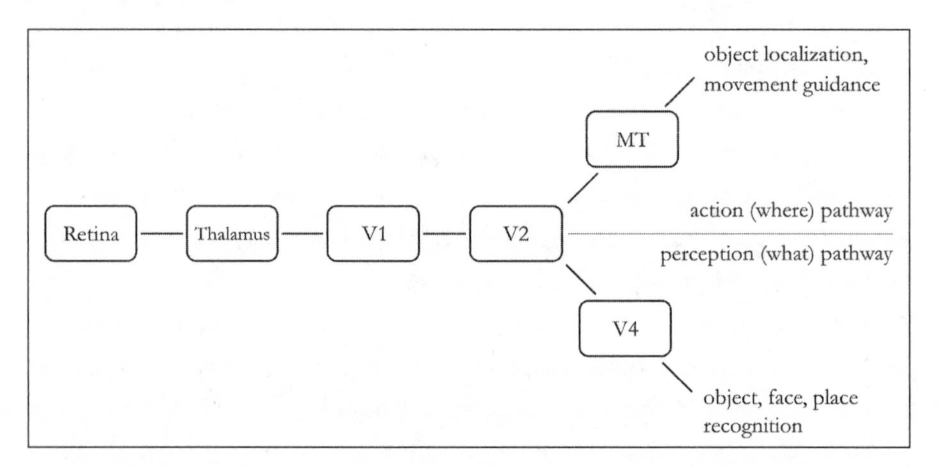

FIGURE 3.7. A highly simplified schematic of the action and perception pathways. Not shown are the many connections, both feedforward and feedback, that exist between the different visual areas.

The primary visual cortex is located at the very back of the brain's cerebral cortex in the occipital lobe. If you were to move from the back of the cortex to the front along the top of the brain, you would transverse in sequence the occipital lobe at the back, the parietal lobe at the center and top, and the frontal lobe in the front. A fourth lobe, the temporal lobe, is located below the frontal and parietal and in front of the occipital lobes. While visual perception is the main function of the occipital cortex, visual processing does not end there. For example, areas of the temporal cortex are important for object, face, and place recognition and areas of the parietal lobe integrate input from vision, audition, and touch and are important for our spatial sense and navigation. So the "perception" pathway involves areas of the temporal cortex, while the "action" pathway involves areas of the parietal cortex. Since the frontal cortex regulates our movements, which are usually visually guided, visual pathways exist throughout the four lobes of the cortex.

As we move up the visual hierarchy from V1 to V2 to higher areas in the occipital lobe and beyond, the receptive field properties of neurons change.[25] Since cells from lower areas converge onto neurons in higher areas, the cells in higher areas respond to information from a wider area of the retina and visual field. As a result, higher-area neurons have larger receptive fields that are less topographically precise. These cells also respond to more complex stimuli. While V1 neurons are stimulated by bars of light in a particular orientation, neurons in different and higher areas of the "perception" pathway respond to whole objects, body parts, faces, or places. Neurons in V4, an area lying between V1 and the higher object-recognition areas, respond best to stimuli midway in complexity between lines and whole objects, that is, to contours and shapes. Liam and Dr. P., "the man who mistook his wife for a hat," could easily recognize geometric shapes, such as triangles and squares, but had a harder time recognizing actual objects. It's possible that for Liam and Dr. P., the intermediate visual area V4, which can recognize shapes, was functioning normally, but the higher areas necessary for recognizing whole objects were not.

Given this visual hierarchy, it seems that we construct our visual world by combining elemental features, such as contours and colors, into shapes and then into meaningful objects. This is what Liam does when he consciously assigns the various lines that he sees with the contours of specific objects. Yet his analytical method is very different (and may use different brain circuits) from how most of us see. We aren't aware of assembling features into whole objects. When we look upon a new view, we instantly grasp the gist of the scene. We see the major landmarks and objects and assign them to basic categories—mountains, trees, houses, tables, chairs. What we don't immediately see are the fine features and details. If we want to see the details, we must turn our attention and eyes toward them. In an analogous way, we recognize printed words without attending to each letter. We recognize a tune without dissecting all the notes. As vision scientists Shaul Hochstein and Merav Ahissar write, "Since the whole is surely built of its parts, how is it that the parts remain unknown, while the whole becomes accessible?"[26]

Perhaps this way of perceiving isn't so surprising because it may describe our first views as infants. A newborn's visual acuity is much poorer than that of an adult. So, what a newborn sees best is a fairly large object that stands out from its background. Even if its features cannot be seen, an object that is moving as a unit against a stationary background is easily perceived. During the first years of life, through their own exploration, infants develop additional strategies to visually segregate objects from their surroundings.[27] From the start, we attempt to perceive whole units, even when we do not know the function of the units or their names.

Through their studies of human visual perception, Hochstein and Ahissar developed the reverse hierarchy theory.[28] With our first glance, visual information travels swiftly from the retina to the thalamus, to V1, V2, and then higher visual areas in a feedforward sweep. Our first conscious sights may develop after activity in the higher visual areas, giving us a view and awareness not of edges, contours, and shapes but of meaningful objects and landscapes. Although we may believe that we have seen most of the details of a scene, psychological tests indicate that we have not.[29] To retrieve most of the details, we need to go back and access the information provided by the lower visual areas.

Information along the visual pathway doesn't go in one direction only. The higher visual areas talk back to the lower ones. Indeed, there are massive feedback connections from higher to lower visual areas. Continuous back-and-forth dialogs occur between neurons and networks at all levels of the visual hierarchy. If I tried to draw all these feedforward and feedback connections as arrows in the diagram in Figure 3.7, I'd end up with an indecipherable web.

We are not born knowing what a chair or a dog looks like. This requires the experience of seeing chairs and dogs in different combinations of light and shadow and from many different viewpoints. As we learn to recognize objects and assign them to basic categories (tables, chairs, houses, dogs, etc.), new circuits and networks develop in our higher visual areas, and new pathways, going in both directions, form between lower and higher regions. That Liam, after his IOLs, saw long, continuous lines everywhere suggests that he had functioning orientation-sensitive V1 neurons and that long-range connections existed between them. The fact that the lines he saw did not automatically suggest the contours of discrete objects may indicate that his higher object-recognition areas were poorly developed.

NORMALLY SIGHTED PEOPLE HAVE STRATEGIES, LARGELY UNCONSCIOUS and automatic, to interpret what they see, and Liam makes use of some of them. As we take in a scene, for example, we do not treat all parts of the visual world equally. We segment the scene, focusing on solid objects, usually those in the foreground, and pay less attention to background details. In other words, we divide the world into figure and ground, and each part has its place in the three-dimensional layout.

Danish psychologist Edgar Rubin was the first vision scientist to define the study of figure and ground, which he introduced in his doctoral dissertation in 1915. He cut nonsense shapes out of cardboard and projected them onto a cloth screen, often instructing his subjects to consider one part of the display as the figure and the other part as the ground. These experiments may sound trivial, but they were not. Rubin noted that people experienced the figure quite differently from the ground. Figure and ground were seen on separate planes, with the figure tending to stand out in front, thus occluding the ground. Lines and contours in the display were seen as outlining the figure, not as part of the ground. The figure had the property of "thingness," while the ground faded into a formless, substance-like surround. People also remembered the figure much better than the ground.[30]

Rubin is most known for the famous face / vase illusion (also called Rubin's vase), which he created for his doctoral thesis (Figure 3.8). If you look at this illusion for some time, your perception alternates from seeing a white vase against a black background to seeing two black faces against a white background. When the shift happens, you have switched your interpretation of what is figure (the vase versus the faces) and what is ground. The figure seems to come to the fore while the ground recedes into a formless surround.

FIGURE 3.8. Rubin's face / vase illusion. A vase or two faces?

Rubin's vase is a classic example of an ambiguous figure whose ambiguity results from how we assign the figure's borders. If we see the vase, then we have assigned the borders between the black and white areas of the illusion to the vase. When we switch and see the faces, we have switched how we assign the borders, which now outline the faces. We cannot see both faces and vase at the same time. If we could, this would present a contradiction. When we assign the border to the vase, then the vase is located in the front of the picture. When we assign the borders to the faces, the faces stand out in front. If we saw both the vase and the faces at the same time, then the vase would stand out in front of the faces, which would stand out in front of the vase—an impossibility.

One role of higher-visual-area neurons may be to coordinate the firing of networks of V1 neurons that respond to a scene's figure as opposed to the ground.[31] As we look at an ambiguous figure and experience a change in perception, this coordination may change. When I showed Liam an ambiguous figure called the Wilson illusion, he could see the two images alternating, one of a man wearing a bulky coat and the second of a face (Figure 3.9). That Liam was able to make this perceptual switch with the Wilson illusion suggests that flexible communication had developed between V1 and some higher visual areas.

FIGURE 3.9. Wilson's illusion. A man or a face?

However, a real-life visual scene usually contains far more than one figure against a formless background. Many objects appear in the scene, and some objects partially block our view of others. In the early twentieth century in Austria and Germany, a new school of psychology, Gestalt psychology, developed. Gestalt psychologists suggested that we don't see each individual feature of a scene in a point-by-point manner. Instead, we group things into perceptual wholes, and this grouping occurs automatically, or without conscious thought. They considered this grouping to be a basic, organizational property of our visual system and proposed several Gestalt principles that are automatically used to organize a scene and differentiate objects within it.[32]

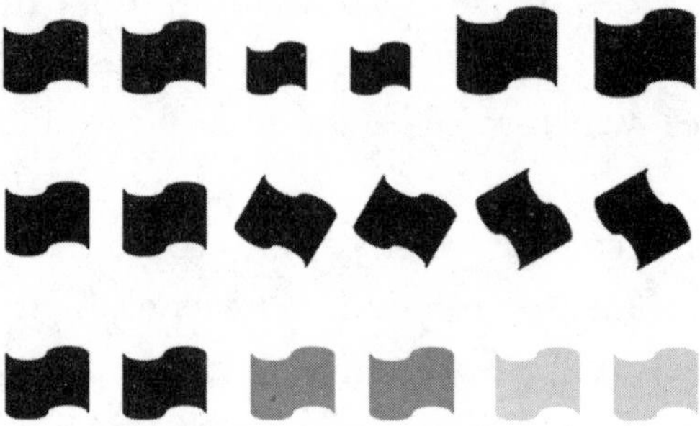

FIGURE 3.10. Examples of Gestalt groupings.

In Figure 3.10, for example, we instantly see the shapes separated into pairs organized by various visual features according to the Gestalt principle of similarity (size in the top row, orientation in the middle row, and grayscale in the bottom row). According to the Gestalt principle of closure, features enclosed by a boundary will be seen as grouped together, and according to the principle of continuity, features that have a similar orientation and form a continuous line will group together as a perceptual whole. With a moment's concentration, you can see the circle "pop out" in Figure 3.11.

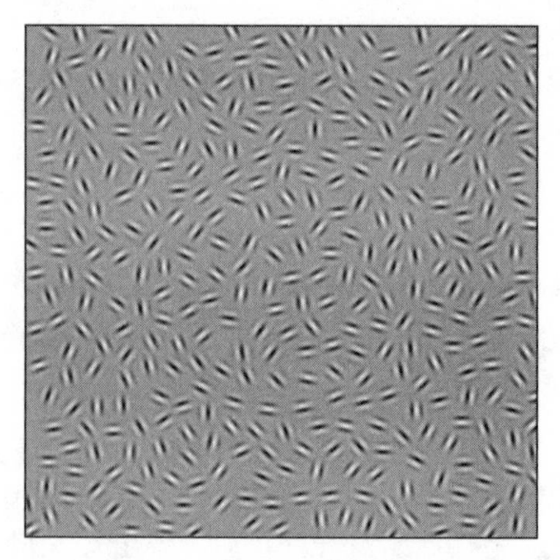

FIGURE 3.11. Contour integration: Do you see the circle?

Camouflage works in large part by taking advantage of Gestalt principles. Various parts of the animal become fragmented and associated not with the rest of its body but with its surroundings, making the animal hard to see. In Figure 3.12, we confuse the pattern on the skin of the copperhead snake with the leaves and ground around it. As a result, we may miss seeing the snake altogether.

FIGURE 3.12. Can you find the snake in the leaves?

Gestalt theories of perception have been criticized for being descriptive and somewhat vague rather than providing a plausible and testable mechanism for how perception works. Yet Gestalt psychologists do describe what we see. What's more, studies of visual nerve cells suggest that some neurons respond to Gestalt groupings and that neurons in higher visual areas may coordinate the activity of lower-area neurons to respond to these groupings.[33]

It seems to me that Liam uses Gestalt principles to find objects in a scene. This process for him is not always automatic, however, but requires careful analysis. "I think of a line as the difference between two colors, the point where two colors meet," Liam wrote. According to Gestalt rules, Liam uses the principle of similarity to group each color region in the scene and then uses the principle of continuity to discern the line. As Liam first consciously and then automatically recognizes Gestalt groupings, new pathways and networks may be forming between the different visual areas in his brain.

In 2012, seven years after his surgeries, Liam drew the picture in Figure 3.13 for Dr. Tychsen. I loved this picture, bursting as it does with bright colors (not seen in this grayscale reproduction) and bold forms and used it as my screen saver for my computer at work. His picture also suggests an organization along many Gestalt principles. There are regions organized by similar shapes. Other shapes are

FIGURE 3.13. Liam's abstract drawing.

enclosed by bold outlines, reflecting the principle of closure. (See, for example, the circles and rectangles separated by dark, thick outlines on the right edge of the picture. Some rectangles are nested within larger rectangles within even larger ones.) Other shapes form continuous lines, illustrating the principle of continuity (see, for example, the line of squares coursing through the middle of the picture).

The bold lines and colors in Liam's picture suggest real things. When he first emailed me the picture he drew for Dr. T., I described in detail some of the images I saw: two circles connected by a curvy line looked like eyes and glasses, and a triangle in the middle next to a straight line looked like a flag on a pole. Liam wrote back to me describing how others too saw real things in his drawing. "I don't see any of it," he said. "I see what is literally drawn on the page." With his usual good humor, he added, "There's supposedly a dragon in there somewhere."

We use perceptual groupings, such as those that create contours or shapes, to help create object categories. A folding chair and a recliner chair look very different, but we instantly pick out their common properties and categorize them both as chairs. A Chihuahua and German shepherd are easy to distinguish from one another; yet we have no trouble assigning them both to the same "dog" category. This ability to discover and extract the common properties among different members of a group results in perceptual learning. The patterns we extract may be represented by networks of interconnected neurons located throughout the brain, and these networks develop with experience and in response to the objects encountered most often.[34] As Liam took in new sights, he had to use his other senses and analytical skills to assign these novel visual sensations to familiar categories, thereby building new networks between visual and other sensory areas.

Objects belong to landscapes, and landscapes have a certain degree of predictability. In one brain imaging experiment, people were shown photographs of various scenes, such as forests, beaches, or industrial landscapes.[35] One quarter of each scene was occluded with a white patch. As a result, V1 and V2 neurons responding to the whited-out section didn't receive any input from the eyes. Yet, they showed activity that correlated with the nonoccluded parts of the scene. This activity resulted, in large part, from input from higher visual areas, areas that provide an overall or global sense of the scene.

When participants were asked to make line drawings of the whited-out part of the landscape photographs, they could accurately fill in the missing information. Indeed, these drawings suggest that they created an internal model of what was missing in the occluded area. This model may develop in the higher visual areas and, via feedback connections, influence the activity of lower-area neurons. In real life, we may continually make and revise such mental models to anticipate what is present beyond what we immediately see.

Liam struggles to interpret pictures, and he likes pictures of landscapes least of all. Without the childhood experience of seeing far and wide, he cannot create good internal models of the expanded landscape. Neurons in his early visual areas may respond to features of the landscape, but without mental models of the larger scene, his visual neurons and networks are not tuned to the most relevant identifying structures and regularities in the landscape. Especially for distant views, Liam struggles to make sense of what he sees.

I was reminded of the importance of internal models to help organize our visual world when my students first viewed the internal arrangement of plant leaves. Leaves may appear flat and thin, but they have a beautiful, ordered, three-dimensional structure. From past years, I knew what would happen. I showed the students how to peel off the outer covering of a leaf and then look at their sample under the microscope. When I asked them what they saw, they replied, "I see a lot of green." Nothing made sense to them. Then we examined plastic, three-dimensional models of leaves as well as prepared slides on which the different internal leaf structures were stained different colors. When the students returned to their original leaf sample, they no longer saw a sea of green. Instead, new structures emerged. They were able to see the whole intricate arrangement—the way the outer cells fit together like a jigsaw puzzle, the arrangement of the inner cells, and the internal air spaces and surface pores that allow for exchange of carbon dioxide and oxygen. I was particularly excited by how, in a matter of hours, they could see so much more than a jumble of green. Here was a nice example of perceptual learning.[36] The raw data coming from my students' eyes did not change, but, through training and attention, they were able to extract much more information from them and assign meaning to what they saw. As I watched my students discover the leaves, I wondered what sort of connections between different visual areas had changed in their brain.

Most of the students forgot the details of leaf structure over the ensuing months; the information was not relevant to their daily lives. But those with good visual memory and imagery may have retained the information for longer. For Liam to recognize objects from day to day, he had to develop his visual memory and imagery, powers that were weak from years of near blindness.

Expectation too plays an enormous role in what we see. This idea was driven home to me a few years ago during a casual glance out my kitchen window at the birdfeeder outside. Normally, the feeder is frequented by small birds, such as chickadees and finches, which I identify in an instant. But at that moment, five large wild turkeys were standing around the feeder, looking through the window directly at me. Their appearance was so bizarre and unexpected that I struggled to figure out the scene. Even though the turkeys were much larger and more distinctive than the chickadees and finches, it took me a lot longer to recognize them because I didn't expect to see them. As I initially turned my head to look outside the kitchen window, my visual system may have readied itself to pick up the features of the smaller birds, thus creating momentary confusion when wild turkeys appeared instead.

These stories emphasize that our vision results from a combination of "bottom-up" and "top-down" processing. "Bottom-up" implies constructing the visual world from the smallest pieces of visual information. It depends on the input coming into the lower visual areas. But we cannot think of visual neurons as responding only to the "bottom-up" stimuli that our eyes see. Their activity is modified by input coming from their neighbors, from other levels in the visual hierarchy, and from other regions of the brain. Prior experience, past associations, and attention to the important targets for the task at hand all influence the firing of lower-area neurons, a process that depends upon feedback from higher visual areas and is therefore called "top-down." Since we all have different experiences, needs, and desires, "top-down" influences differ from person to person. We all see the world through our own perceptual lens.

"Up close," Liam wrote, "things are more like objects than visual chaos, and there is a definite difference when I see something further away, it has no meaning, and I struggle to tell if a bar of color is the front of a truck or side of a bus or roof of a building. If people even stand slightly further away and talk to me or say hi from down the hall, it has a very different feeling, and it doesn't seem as real." After his IOL surgeries, Liam's eyes provided his V1 neurons with the input they had always been waiting for. But he lacked the visual experience of seeing beyond a few feet, so he had not yet developed the "top-down" processing that organizes these local details into coherent objects and landscapes. He had to rely heavily on "bottom-up" processing and consciously piece together the visual world from its parts. This was particularly hard to do when looking in the distance where fewer details can be seen.

It is almost impossible for most of us to look out upon a scene and see lines and colors without also seeing the objects they make up. So it is hard to imagine what Liam saw. Although each of us perceives the world differently, I can safely assume that the objects I recognize at any given moment, people near me see too. These objects fall into basic categories that we all understand. If I see a chair or a dog, so do they. But envisioning Liam's new perceptual world, a fragmented scene full of disembodied features, is an entirely different matter. I remember how frustrating it was for me to describe my new 3D views after gaining stereovision. Those who always had stereovision usually took it for granted and didn't know what they had, while those who had always been stereoblind couldn't imagine what they were missing. How much more difficult it was for Liam to describe his strange new world.

But neuronal connections, like ruts in the road, deepen with use. As Liam began to recognize individual objects among the confusion of lines, he may have developed new networks in higher object-recognition areas.[37] In addition, neurons in higher visual areas may have selectively facilitated the firing of lower visual neurons that respond just to those objects as well as inhibited the activity of neurons that responded to the background. With each day, Liam's vision became more "top-down," providing more meaning to his visual world.

Nowhere is our ability to see the whole from the parts more evident than in our ability to recognize faces. Most of us can identify hundreds of different faces, even those we have not seen in years. Indeed, our ability to recognize faces is among our most impressive visual skills. Before receiving his IOLs, Liam couldn't see individual facial features. Now that he could see more clearly, could he recognize a person by his or her face alone?

– chapter 4 –

Faces

GIUSEPPE ARCIMBOLDO WAS A RENAISSANCE PAINTER WHO CREated bizarre portraits of people whose head parts were made up of vegetables and tree roots. Although his conventional religious paintings have long been forgotten, his vegetable portraits are still popular. In addition to being fun to look at, these paintings tell us something about our face recognition skills. In Arcimboldo's painting in Figure 4.1, we see a head with a bulbous nose and round cheeks. Turn this picture upside down, and the face vanishes, appearing instead as a bowl of vegetables.

The parts of the face in the upright painting are not actual facial features. The nose is a carrot, and the cheeks are onions. But we instantly see the whole picture as a portrait. We have gone beyond our view of the individual vegetable parts and assembled them into a human face. Indeed, it is much harder to recognize a real face (not just Arcimboldo's vegetable face) when the image is upside down because the arrangement of the parts is all wrong.[1]

Artist Chuck Close experimented with a similar idea in his famous, much-larger-than-life portraits. He would photograph a subject, draw

FIGURE 4.1. *Portrait with Vegetables* by Giuseppe Arcimboldo.

a grid on the photo and on his canvas, and then fill in each square on the canvas grid. Strikingly, each square was filled with a design that didn't match the details of the face at all. But the distribution of lights and darks on the canvas followed the shading of the person's face in the photo. If you looked at the paintings close up, you saw an abstract design made up of colorful little squares. If you looked from a distance, however, all the squares merged together to produce a gigantic face. As you approached the canvas from a few yards away, you could experience the transition from a holistic view of a face to an abstract design made up of individual squares.[2]

Paintings by Arcimboldo and Close teach us that we recognize faces in a holistic way as opposed to scrutinizing every detail. This makes ecological sense since the details of a person's face change constantly. We still recognize a person when they frown or smile, get a new haircut, grow or shave off their beard, change their hair color, take on or off their glasses, and so on. We may recognize a person whom we haven't seen for years, even if during that time they have grown from child to adult.

People vary greatly in their face recognition ability, and some have extraordinary skill. Recently, I went to a baby shower where we all

FIGURE 4.2. Chuck Close, *Self-Portrait*, 2007.

had to bring photographs of ourselves as infants. Since most of us were born in the 1950s, these photographs were small black-and-white snapshots, and our faces were no bigger than a thumbprint. The photos were placed on a big board, and we all had to guess whose baby picture belonged to whom. I call one of my friends "the woman with two brains" because she has an uncanny ability to recognize, with a fleeting glance, a face that she hasn't seen in years (as well as to recall all the biographical details of that person's life). At the baby shower, this friend instantly identified all our baby photos. Since our faces had changed in considerable detail over the ensuing sixty or so years, she must have recognized our faces holistically.

Like many others who gain vision in adult life, Liam concentrated on details but had trouble assembling them into meaningful objects, and this problem made it very difficult to understand faces. What's more, he didn't describe his first view of human faces as beautiful or inspiring. Right after the first surgery, Liam was disgusted when he saw the way his mother's mouth moved when she talked. Prior to receiving his intraocular lenses, Liam couldn't see details on another person's face; the nose and mouth were just a blur. He knew that his mouth moved when he talked, but it was a revolting shock to see the details of the red lips and tongue in others.

Indeed, without a holistic sense of a face, it was impossible to recognize people from moment to moment. Their faces transformed entirely when they changed expression or talked. This predicament reminded me of comments by Chuck Close, who despite, or maybe because of a condition called "face blindness" (prosopagnosia), paints portraits. Close cannot recognize faces. As he said at a talk at the World Science Festival in 2010, "In real life if you move your head a half an inch, to me it's a whole new face I've never seen before."[3] The same was true for Sheila Hocken, who had poor sight as a child, became blind as an adult, and recovered her sight after cataract surgery. She wrote in her memoir *Emma and I*, "People don't have one face, they have hundreds."[4] When most people look at a moving face, they see not only the expressions but the person behind the face. Chuck Close and Sheila Hocken, on the other hand, see something different with every facial movement or expression. Every change makes a whole new image. Problems with recognizing faces and facial expressions are very common among people with long-term blindness who gain sight as adults.[5] Even people blinded by cataracts from birth whose sight is restored within the first year show some deficits in face recognition.[6]

Infants, at just nine minutes old, exhibit a preference for looking at a human face.[7] This remarkable fact was discovered during an experiment in which three different pictures were moved across a newborn's field of view. When a face pattern (an oval for the head, enclosing shapes that looked like eyes, a nose, and a mouth) was waved in front of the baby, the child would turn his or her head and eyes to follow the pattern. But if the features were all mixed up so that the pattern no longer resembled a face, the infant tracked the image much less. While we are not born with an innate ability to recognize household or most natural objects, we may be born with a rudimentary face-detecting skill.

What's more, babies only forty-eight hours old prefer to look at their mother's face over the face of other women.[8] It's possible that, within the first two days of life, they are able to match their mother's voice, heard in utero, with their view of her face. While the particular objects infants see depend upon where they live, all babies must interact with other people to survive. It makes sense, then, that infants show a special preference for looking at a human face, particularly the face of their primary caretaker.

One area of the brain that is particularly active when we look at faces is called the fusiform face area.[9] Intriguingly, this same area "lights up" when expert chess players look at a chess board.[10] Why would an area of the brain important for the recognition of faces be activated in chess experts when they view a chess game? To recognize a face, we need to see more than the eyes, nose, and mouth. We must analyze the spatial relationships between all these features. Similarly, an understanding of the spatial relationships between game pieces is crucial for winning at chess. The fusiform face area is good at recognizing global, spatial patterns. Circuitry present at or soon after birth and an infant's preference for looking at faces likely bias the fusiform face area to become a face recognition area. This role is bolstered further by our lifelong experience with viewing faces and the importance of face recognition in everyday life. A chess professional may tap into the global, spatial processing circuitry of the fusiform face area for analyzing chess.[11]

When Liam was in college, he initially enjoyed playing chess with the chess club but eventually gave it up. He kept overlooking possible offensive moves on his side and threatening moves from his opponents. Something analogous happened in real life when he would overlook people and objects right around him, and he didn't like video games in which he had to avoid attacks from all sides. All this made him wonder if he had a poor awareness of his visual periphery and of global, spatial patterns, as well as, perhaps, a poorly developed fusiform face area.

A newborn's visual acuity is much poorer than that of an adult, so very young babies will see a face in a lot less detail. They may find it easier to recognize their mother's face from outer facial features, such as hair style and hair color, than from inner facial features, such as the distance between the nose and mouth.[12] Liam too depended upon outer facial features to recognize his own mother, which is why he never let his mom wear a hat. As babies mature, however, they see more details in a person's face. This did not happen for Liam before his IOLs. Even after the operations, he struggles with face recognition and finds it easier to recognize a face from a photograph or on TV than a face in real life.

Habits may play an important role here too. Since individual features, including the shape of the nose and mouth, and their spatial relations were all a blur to Liam before his intraocular lenses, they gave him very little information. Even if he was born with an innate preference for looking at human faces, his childhood experiences wouldn't have reinforced this habit. In college, Liam, who never shrinks from a challenge, decided to take a class in American Sign Language, a language built on movement, not sound. This class certainly put his visual skills to the test. In one class, the students watched a video of a person signing, and this person had such a bad facial tic that it distracted viewers from following his signs. But Liam did not notice the facial tics. The other students were flabbergasted. As Liam wrote, they demanded, "HOW could you Not notice!?!?" Part of his face recognition problem may stem from the simple fact that he does not automatically study a face.

One of the first people Liam recognized with his new vision was his college professor Joe. Here's what Liam wrote in an email to Dr. Tychsen in 2012: "Joe has black and grey hair and a mustache, and I recognize hair best instead of faces. Especially mixed color hair and facial hair (even though when I look away I can't definitely describe it). So when I recognized him automatically on campus outside of class, I had to go tell him how special it was and that he was the first person I recognized like that."

Liam assigns the way people look to broad categories based on the features that he sees best (short or long hair, glasses or no glasses, etc.). Sometimes, he feels, these categories give him an edge in seeing a resemblance between people. This happened when Liam was watching a video and saw a person in it that looked like Dr. Tychsen. He mentioned this to his mother, who reacted with great relief. Cindy thought that the person in the video looked like somebody she knew, but she couldn't place him. Since the person in the video was much younger than Dr. T., Cindy didn't make the connection. Liam saw it immediately.

Many people with face blindness have a hard time recognizing faces but have no difficulty recognizing facial expressions. With expressions, too, Liam ran into problems. When, eight years after his surgeries, I showed Liam a set of cartoon faces displaying all sorts of emotions, including happy, surprised, skeptical, disapproving, confused, fearful, and sad, Liam told me that the only expressions he really understands are happy and sad. Could this problem result in part from the way Liam looks or doesn't look at faces? I was struck by an article I read about a woman called SM with a very rare disease that destroyed her amygdala, a structure in the forebrain that is involved with our experience of fear.[13] SM does not show normal fear responses but is indiscriminately trusting and friendly. While she is able to sketch a picture of a face expressing happiness, sadness, surprise, anger, and disgust, she cannot draw a face showing fear. Nor can she recognize fear in photographs of human faces.

Most of us judge facial expressions by looking primarily at the eyes. When scientists monitored SM's eye movements as she looked at photos of human faces, they discovered that she didn't direct her gaze at the eyes. It is possible to judge many emotions, such as happiness, by looking at the mouth, but to recognize fear, you must look at the eyes. Indeed, if the eyes are erased from photos of a human face, control subjects lose their ability to recognize fearful expressions. Remarkably, when SM was instructed specifically to look at the eyes in the photos, her ability to recognize fear increased to normal levels. Liam does not have SM's neurological problems. But, like SM, he may not recognize certain facial expressions because, after a childhood of near blindness, he does not have the habit of looking at other people's eyes and faces.

Yet, if Liam were to look directly at your eyes and study your face intently, you may find such scrutiny unnerving. A baby may stare at you, but most adults do not. We generally find a balance, attending to others without looking so intently that they feel we are invading their privacy. One reason Liam may find it easier to recognize a person in a photograph or on TV is because he can study the person in the photograph without that person being aware of the scrutiny.

In her memoir *Emma and I*, Sheila Hocken writes that her face became more mobile, expressive, and alive after she gained sight and saw the expressions on other people's faces.[14] When I first met Liam, he did not look directly at me, hardly showed any expression on his own face, and spoke quietly. But he has since become more animated, with his face breaking into big smiles. Yet being aware of other people's feelings through their expressions means that other people are aware of yours. All this can make a shy person like Liam all the more self-conscious.

When it comes to vision, what matters most to Liam is how well it helps him function in daily life. He is comfortable with his social group in St. Louis now, so does not worry too much about his face recognition skills. Instead, he has concentrated on other ways he can use his vision to live on his own and enrich his life.

– chapter 5 –

Finding Things

IN 2012, WHILE LIAM WAS IN COLLEGE AND LIVING AT HOME IN Columbia, Missouri, his mother and brother left for Oklahoma to take care of his grandmother. So Liam headed to the supermarket to shop for food on his own. "I am horrible at shopping," Liam wrote to Dr. T. "I don't know where I should look, I don't know what [the item] I'm looking for will look like on the shelf, I can't remember what [the item that] I'm looking for looked like when I [previously] had it . . . With produce, I don't know any of them. I can see they are all different colors but at the same time they all look like they are the same color and shape, another contradiction."

Similar problems hound Liam in a cafeteria line where he must distinguish and select different food items. A fruit salad or rice dish is a mixture of shapes and colors. Indeed, when he first enrolled as a student at Washington University for the post-bac program in 2013, he wrote about his "experimentation" in the dining hall. He couldn't read the menus on the walls, yet felt pressured to decide and move quickly lest the cashiers get annoyed. So he was a little malnourished for the first few weeks, helping himself mostly to the salad bar, where he didn't recognize all the selections and ended up eating some mystery foods.

Ever resourceful, Liam found a list of dishes for one cafeteria online. This worked well, unless he accidentally asked for something that was not there. "What you see is what we have," the staff would tell him.

We live in a cluttered world. As with the apples lumped together in a store bin or the mixture of foods in a salad, we usually don't see the border around every object in full. Instead, an object in front may partially block our view of other objects behind it so that we see a complete border only around the object in front. We have to infer the missing parts of the partially hidden objects.

"I think I use a lot of two-dimensional shapes and lines to describe things instead of three-dimensional objects," Liam told me, and his two-dimensional interpretation of a scene will make recognizing whole objects more difficult, particularly in clutter. Indeed, when I showed Liam some images that could be interpreted as three-dimensional, he saw them as flat.

For example, in 2014, I showed Liam the famous Kanisza triangle (Figure 5.1). In this figure, most of us see a bright-white, down-pointing triangle hovering above a second up-pointing triangle. Pac-Man-like figures surround the corners of the down-pointing triangle. While the Pac-Man figures help define the corners of the down-pointing triangle, the rest of the triangle's outlines are not actually present. They are merely suggested, thus forming illusory or subjective contours.[1] Partic-

FIGURE 5.1. Kanisza triangle.

ularly striking about this illusion is that the down-pointing triangle appears to be brighter than and floating above the up-pointing triangle. In other words, the illusion is seen in three dimensions. Liam saw a suggestion of the three corners of the down-pointing triangle, but the triangle disappeared in the middle. It did not appear as a bright triangle that popped out in front of the rest of the figure.

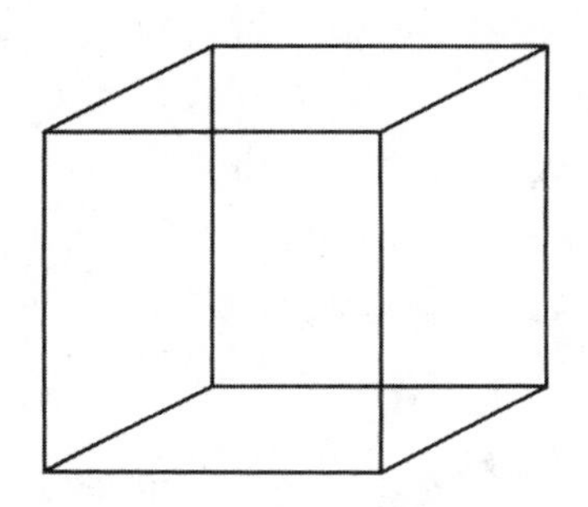

FIGURE 5.2. Necker cube.

In school, Liam learned to draw a Necker cube, which he would do in a very scripted way (Figure 5.2). When he concentrates, he can see the cube as three-dimensional, but, with a casual glance, it looks to him like a flat collection of lines.

Similarly, most of us see the picture in Figure 5.3 as a piece of paper folded vertically in three parts. The left third may appear to be folded toward us while the rightmost part is folded back. Then, the image may shift so that the leftmost part folds back and the rightmost part folds forward. Liam saw the paper flip in the same way, but he also saw the picture in a third way—the whole piece of paper could appear flat with angled edges.

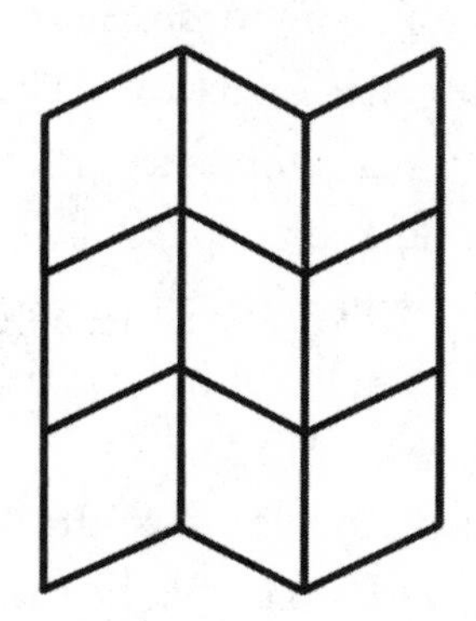

FIGURE 5.3. Which way does the paper fold?

So, I was surprised and impressed when, two years later, I emailed Liam the puzzle in Figure 5.4. In the left part of the figure, we see a lot of shapes that do not immediately coalesce into anything recognizable. But there's something recognizable in the image to the right. If you don't see it, use this hint: the image contains multiple copies of the same letter. Then you may see the *B*s partially occluded by a thick, black network. The parts of the letter *B* that are seen behind the black network in the right half of the figure are exactly the parts represented in the left half. Yet, in the latter, it was very difficult to pick out the *B*s. Ironically, the network of black lines, while partially blocking the letters, helps us to see the fragments as making up the *B*s!

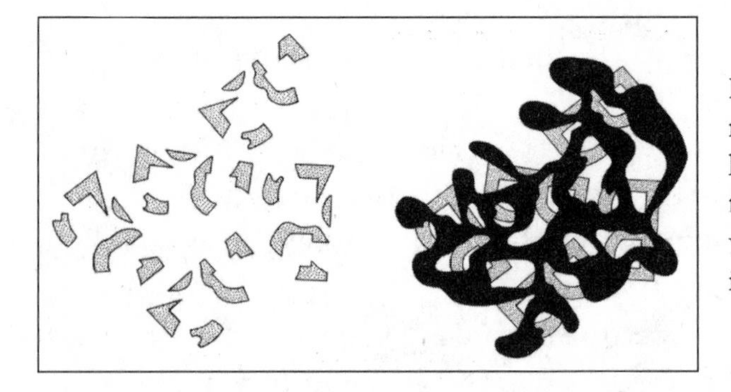

FIGURE 5.4. The fragments we see in the left half of the figure take on new meaning when we view them in the right half.

We see the *B*s in the right but not the left half of the figure by a change in the way we assign border ownership. In the left half, we see pieces of the letters *B*, and the border of each of these pieces appears to belong to the piece itself. Thus, these fragments don't come together to form the *B* shapes but remain as discrete entities on one flat plane. In the right figure, we assign the common borders between the black and stippled regions to the occluding black network. This allows us to recognize the fragments as belonging together, as parts of larger shapes, the letters *B*, which continue behind and are partially obscured by the black lines. By assigning border ownership to the occluding black network of lines, the fragments behind it connect with one another into recognizable letters.[2]

When I emailed Liam the two parts of the above figure with the hidden *B*s and asked him what he saw, he responded that he saw the left part of the figure as made up of pieces but, over time, saw the *B*s emerging on the right. So, Liam could see that the black area represented one surface and the gray parts made up another. After seeing that the borders in the right-hand image belonged to the surface that made up the occluding black net, he could fill in the missing contours of the *B*s. He interpreted the picture as representing a three-dimensional scene. Such thinking should help him recognize objects in the real world, even when they are partially occluded by others.

OF COURSE, IT IS EASIER TO INTERPRET THE WORLD AS THREE-dimensional if you see in three dimensions, that is, with stereovision (stereopsis). Since the surface of an object may approach and recede in depth, recognizing objects may require not only seeing their contours but following them in depth. Stereovision helps to see those depth changes. When I gained stereovision in midlife, I was astonished by how much crisper everything looked.[3] The outlines of objects stood out clearly. Part of Liam's struggles with sorting out fruits and vegetables in the supermarket or recognizing salads in the cafeteria may result from his poor stereovision.

To see with stereovision, you need to aim the two eyes simultaneously at the same place in space and then mentally fuse the images provided by the two eyes. The result is a vivid, three-dimensional view in which you can see not only the solidity of objects but the volumes of space between them. Infants demonstrate the ability to see with stereovision at three to four months of age, and this very early age of onset suggests the importance of stereovision for visual development.[4] In contrast, it is not until six or seven months of age that infants use "pictorial cues," such as perspective and shading, to interpret depth.[5] Indeed, some vision scientists have proposed that stereovision guides the development of these later perceptual skills.[6]

Throughout childhood, Liam saw best through his left eye. His strabismus, or eye misalignment, led to double vision and visual confusion, so his brain suppressed the right eye's input. He had to close his left eye to see anything through his right, but seeing through his right became painful after only a few seconds. As a toddler, he unconsciously developed intermittent right-eye closure.

So it was a shock when in the summer of 2010, five years after his surgery, Liam woke up to find that he was no longer closing his right eye. Liam wrote that his right eye was seeing "against my will." Surprisingly, it was no longer painful to see with the right eye, but Liam found it "horrible." The elimination of the closure had opened up a large swath of his right visual field, and the new visual input was overwhelming. He tried squinting the right eye shut but couldn't keep that up for long. The next day, this new way of seeing was slightly more tolerable, and after a week, he had gotten used to it. Now, he notices that the transition from the left half to the right half of his visual field appears seamless.

Clinical tests in Dr. Tychsen's office showed that Liam could fuse images from the two eyes into one and see the fused image in stereo depth, but his stereoacuity (three thousand arcseconds), a measure of how far apart two targets have to be to see the depth difference, is poorer than normal. Since Liam's eyes were misaligned in infancy and since he has some eye-to-brain misrouting due to albinism, his ability to fuse images and see with any stereovision is impressive. Curious about his own stereovision, Liam looked at the stereo pair in Figure 5.5, turned in his eyes, and fused the images into one. When they were fused, he saw the inner circle not off to the right or left but in the center of the outer

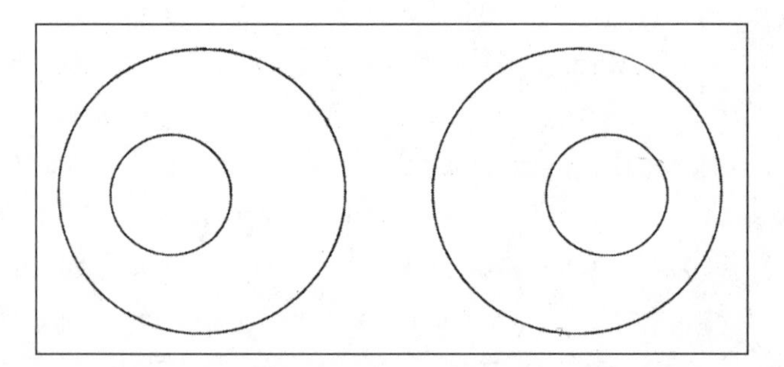

FIGURE 5.5. If you fuse these two images by turning in your eyes, the inner circle should appear to pop out in front of the outer circle. If you fuse by looking through or beyond the page, the inner circle should appear behind the outer circle.

one as it should appear. However, he did not see the inner and outer circles at different depths.

So, when I visited with Liam in June 2014, I brought with me the Quoits vectogram, which consists of two polarizing sheets, each with an image of a rope circle (Figure 5.6). When you put on the polarizing glasses (similar to the ones used for 3D movies), each eye sees only one of the rope circles. If you slide the rope circles apart in the horizontal direction and then, wearing the polarizing glasses, fuse the two images, you see one rope circle that appears to be floating in space. With some effort, Liam was able to see the rope circle hover in front of the sheets. Since fusion takes effort, he may not be fusing on a regular basis. Like me, he might need vision training to develop the fusion habit in order to automatically see in 3D.

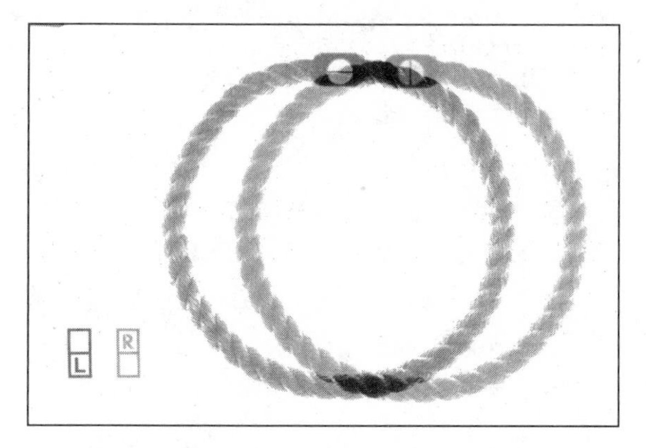

FIGURE 5.6. The Quoits vectogram.

Marius von Senden, in his book *Space and Sight*, reported that people who had been blinded when young and then gained sight in both eyes recovered vision more readily than those who gained sight in only one eye.[7] Liam has only a marginal ability to use the two eyes together for stereovision. So, unlike an infant just beginning to see, he has to learn to interpret a scene and segregate the objects within it, like the apples in a grocery bin, without this powerful cue.

When I saw Liam in St. Louis in 2014, I witnessed his strategies for getting through a supermarket or superstore. Now, nine years after the surgeries, he was living alone in his own apartment, but his mom, Cindy, came up from Columbia for the day. The first place we went was Walmart, where Liam planned to pick up a tube for a flat bike tire and show us how he shops in such a huge and busy store. My first thought, upon entering the brightly lit building, was how resilient Liam is. He had to tolerate the store's fluorescent lights, where glare could be a problem. In his own apartment, he keeps the shades drawn and is most comfortable in dim light.

Always logical and analytical, Liam recounted his strategies for shopping at Walmart. He said people in St. Louis were more aggressive than in his hometown, and this applied not only to the way they drove but also to the way they shopped. In stores, Liam would park his cart in quiet aisles, like the pet aisle, and then search for an item someplace else in the store. It was easier to get around people without the cart.

In the frozen food aisle, we looked at the Lean Cuisine packages, one package after another. Liam looked at the words, not at the pictures, because the pictures didn't make much sense to him. He could read the big signs above the aisles when I pointed them out, but he rarely noticed or used them on his own. I took off my glasses, which reduced my acuity to about Liam's eyesight, and the letters on the signs were a bit blurry but still readable.

We went to an aisle with rows and rows of Tostitos salsa jars. The packaging of the medium and mild salsa jars looked almost the same except for the color of the stripe on the lid. The words "medium" and "mild" were written in small font on the middle of the jar. Liam didn't notice the colored stripes, so he had a hard time noting the difference between the two salsas. This problem extended beyond Tostitos jars. The packaging on many items is visually complex and overwhelming in detail. No wonder Liam often came home with the wrong version or size of things or ended up with five of everything because he didn't notice what he already had at home.

Then we went to the pharmacy section. Liam led Cindy and me there. Although this was not the Walmart he normally goes to, all Walmarts are arranged similarly, so he reasoned out where the pharmacy was. He used the nearby cosmetic aisle as a landmark since it stood at a forty-five-degree angle to the other aisles.

We got to the pharmacy aisle and started looking at packages of Mucinex. As with the Lean Cuisine boxes and the Tostitos salsa bottles, I was struck by how the packages and bottles looked almost the same, except for their color. Liam did not mention their color but leaned in close to read the labels, which were more distinctive than those on the Tostitos jars.

NOT ALL VISUAL SEARCH TASKS ARE FRUSTRATING FOR LIAM, however. He can also have fun, as occurred on the afternoon Liam and I spent outdoors in Forest Park, an oasis of green in the middle of St. Louis. As we approached the park, we walked over a wet sidewalk, where the cracks held the water so that they appeared darker than the surrounding cement. To Liam, these dark areas looked like low rails that needed stepping over. The play of light and dark and the appearance of lines, Liam remarked, could sometimes be "lies."

At the park, we went "geocaching." People all over the world use GPS to hide and seek containers, or "caches." Liam loves these sorts of puzzles and treasure hunts, a real blessing given that his whole visual world can sometimes seem like a riddle. To find the caches, which were packaged in small plastic or metal containers, Liam downloaded onto his portable GPS their published coordinates and a map. Two of the caches were hidden in trees and one in a toy rat in the underbrush. Liam was the first to find two of them. When he spotted the geocache in a mulberry tree, he commented that it was "in plain sight." When I needed to stop to tie my shoe, he pointed out a park bench some distance away. In the calm, spacious surroundings of Forest Park, Liam made good use of his vision.

Why could Liam pick out the geocaches on our trip to Forest Park when he had so much more trouble finding items in a supermarket or Walmart? Although, in Forest Park, there were plenty of things to see—trees, benches, paths—these were large items that, with his new eyesight, he could easily recognize and lump into basic categories.[8] While looking for the geocaches, he didn't need to know the particular species of tree or style of bench that we passed. As the reverse hierarchy theory explains, for grasping the gist of the scene, the fine details don't matter.[9] Liam merely had to sense the local layout so that he could avoid bumping into things and then search for the geocaches, which, housed in small plastic or metal containers, stood out from their natural surroundings.

In a superstore, however, every shelf presents a multitude of items, boxed in packages that are stacked like bricks, one next to or on top of another. If Liam wanted to find his favorite salsa, he had to search for a particular brand and then a particular level of spiciness within that brand. His favorite salsa was a member of a category within a category within a category. Since all the different subcategories of salsa jars were arranged next to each other and had similar packaging, the narrower the subcategory he wanted to find, the tougher the search.

Liam's recently acquired sight allows him to see the words on the labels clearly and to see colors, shapes, and object borders as well. However, Liam ignored the colors and designs on the labels. He has a hard time interpreting pictures, so he depends heavily on reading labels, a skill that he mastered in childhood. Being able to read print is a huge advantage for Liam. Many sight-recovery people, who were blind from birth or lost their vision in early childhood, never learn to read print. Once sighted, they are able to recognize letters, particularly capital letters, by sight, but stringing letters together into words is far more difficult.[10] Liam's early vision allowed him to learn to read print, and although reading in high school was arduous, he wisely revived and sharpened his reading skills once he received his intraocular lenses. When I thought of all the signs and labels we encounter every day and how disoriented I felt on a trip to Japan, where I didn't understand the writing, I realized just how critical Liam's reading skills are to him in finding his way, not only in a store but throughout our lettered environment in general.

– chapter 6 –

Vision's Greatest Teacher

BEFORE LIAM'S INTRAOCULAR LENS SURGERIES, IT TOOK SO LONG for him to focus on an object that he could never see it in motion. But after his surgeries, he not only saw moving things but could gauge their direction. The same is true for others who gain vision in adulthood. When Oliver Sacks visited Virgil, whose sight had been restored only weeks earlier, Sacks was impressed with Virgil's extreme sensitivity to motion. Virgil's eyes followed moving objects even though he couldn't identify them.[1] In his book about Michael May, who was blinded at age three and recovered sight at age forty-six, Robert Kurson relates a moving passage about May playing catch with his five-year-old son.[2] It was only a day after May gained sight; yet he could see and track the moving ball and kick it or catch it, even while on the run.

Indeed, as soon as Liam recovered from his IOL surgeries, he and Cindy headed out to the driveway to play catch. Cindy bounced the ball, and Liam attempted to catch it on the first bounce. With the first successful catch, they did a little victory dance on the driveway. Cindy broke into a huge smile as she told me this story; it was the first time her child could use his eyes to track and catch a ball. Although the winter that Liam received his IOLs was cold and icy, he insisted that he and his mom head out to the streets so he could practice kicking a soccer ball. He even made it onto the high school C soccer team. It took some time for Cindy to realize that Liam was using soccer as a way to train his eyes and brain.

When we planned my visit in 2014, nine years after his surgery, I wanted Liam to show me what he enjoyed most with his new vision, and sports was at the top of the list. After our morning trip to Walmart, Liam and I headed off on bikes to an athletic field to play catch with a tennis ball. Liam caught the ball easily and threw it accurately. I threw the ball high in the air, which makes it harder to catch, but Liam caught it anyway. When we rolled and bounced grounders at each other, Liam at first tried to grab the ball by putting his hand on top of it, but after a few throws, he put his hand under the ball to scoop it up. Then I would throw the ball when Liam was running. On the first and second attempt, he slipped as he tried to grab the ball, but after that, he caught it and did not slip. This was something he wanted to practice because it was similar to chasing a hockey puck, and Liam had joined an ice hockey team.

Tired of catch, Liam led me to an outside ping-pong table. We were evenly matched. Liam's forehand was weaker than his backhand. With the forehand, Liam explained to me in his typical analytical way, his paddle was out to the side. But with the backhand, his whole body was behind the ball when he hit it. Liam aimed the ball at the corners, particularly to my forehand. He was aware of where I was and where then to place the ball.

Seeing objects in motion provided Liam with several surprises and epiphanies. Liam wrote, "When I first starting seeing, I could play catch but not do much else yet. Someone threw a white ball to me and, on the first throw, all I saw was this circle, and [as] I was trying to figure out why it was growing, the ball hit me while I was just standing there thinking: 'Oh! I'm catching the white circle thing, ok got it.' From the next throw on, I caught it perfectly and had no trouble." With that game of catch, Liam realized that balls (and other things) loom larger as they approach. Spherical balls (as opposed, for example, to American footballs), he discovered, appear the same from all perspectives, and this makes them easier to catch.

Two years before my visit, Liam had written to Dr. Tychsen, "I love playing sports and it is easy for me visually, no anxiety, no calculations, just seeing." At first blush, this seems surprising. Wouldn't it be harder to see and then grab something that is moving than to take your time and see something that is standing still? But motion, like color, is a visual primitive—a visual quality that can be recognized without any prior experience. We may be sensitive to motion from the moment of birth and sensitive to the direction of motion as early as six to eight weeks of age. Direction sensitivity may depend upon maturation of the visual cortex. Neurons in visual cortical area V1 communicate with neurons in visual cortical area MT, and the firing of MT neurons depends upon the location, direction, and speed of motion on the retina. So maturation of this pathway may lead to direction sensitivity.[3]

OUR ABILITY TO SEE THINGS IN MOTION HELPS SHAPE OUR WHOLE concept of objecthood. An object is easy to distinguish from a stationary background if it moves. Indeed, infants as young as four months old perceive the whole of an object even when part of it is occluded if the parts that they do see move together.[4] Gestalt psychologists call this property—that an object moves as a whole—the "law of common fate."[5]

Since most of us can look at a photograph and recognize the pictured objects, it is easy to think that we see the world like we see a photograph: as a snapshot in a single instant of time. But in real life, we and much of the rest of the world are in motion. Sensing our own movements as well as the motion of other things may be critical to our understanding of time and space. While watching an object move or while moving ourselves, we learn that it takes time to cross an expanse of space. As psychologist Barbara Tversky has written in

Mind in Motion, our thinking may be shaped by the way we see and organize space and move within it.[6]

Indeed, sensing our own motion is critical for visual and cognitive development. Vision scientist James Gibson points out that objects to our right go out of sight when we turn our heads to the left but then come back into sight as we turn back to the right.[7] These objects don't go out of existence—only out of sight. I was reminded of Gibson's thoughts when my granddaughter was about four months old. While I was holding her upright, she would take a long look in one direction, then turn her head and gaze for some time in the new direction before turning back again. She did this over and over again, which led me to wonder if she was testing whether or not things went out of existence when she turned away. Perhaps this is why all sighted babies love to play peek-a-boo.

Seeing and being in motion also teach us about causality. We push something and see it fall. The wind blows, and we see leaves tremble. Psychologist Albert Michotte studied causality with experiments that were as elegant as they were simple.[8] He showed people a display of a square moving in a horizontal line and then stopping when it contacted a second square. If the second square moved as soon as it was touched, his observers reported that the first square caused the second to move. Timing here was critical. If the first square contacted the second square, but then there was a pause before the second square moved, observers believed that the second square moved on its own. We begin using the movement of objects in this way to determine causation early in life, by six or seven months of age.[9]

Liam's fortuitous love of motion was a great catalyst for his own vision development because movement helps with object recognition. Real objects are three-dimensional and usually opaque. As a result, each object occludes part of itself. The front of an object, for example, blocks our view of the back. So we cannot see all parts of an object in a single glance from a single vantage point.

Figure 6.1 shows three images of a chair as seen from three different viewing angles. The left-hand image shows only three chair legs because the view of one leg is blocked by parts of the chair in front of it. Yet, we know that the chair has four legs. In none of the three images of the chair do we actually see the full extent of the chair seat. Only if we looked at the chair from above or below would we see its full length and width. When we stand in front of the chair, as drawn in the left image, the seat appears foreshortened, although its side-to-side extent is seen in its entirety. In contrast, if we move to the chair's side, we see, as with the right image, the full extent of the front-to-back dimension but not the side-to-side. What was width has now become depth and vice versa.

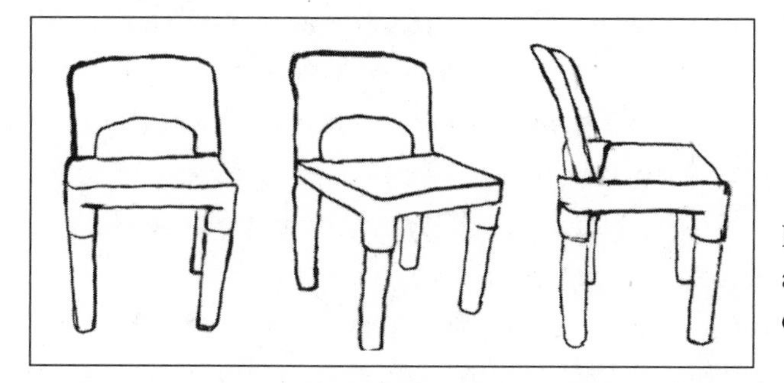

FIGURE 6.1. A chair as seen from three different viewpoints.

All these changing perspectives did not necessarily make it harder for Liam to recognize a chair. Early on, he realized that he could identify objects more readily if he walked around them. As Liam moved around a chair, the chair's appearance would transform smoothly. Figure 6.1 shows three snapshots of what he would see as he circled a chair in a counterclockwise direction, and the way the chair's appearance changed would tell him about its three-dimensional structure.[10] Other cases of sight restoration after childhood report the same benefit of deriving three-dimensional structure from motion.[11]

A fundamental understanding of the solid, three-dimensional nature of objects emerges very early in life. Indeed, in one study of visual development, fourteen- to twenty-week-old infants were shown movies of a three-dimensional object as it rotated about a particular axis. They could then recognize the same object when they saw it rotate about a different axis, and they could distinguish it from other rotating items. If, instead of the movie, the infants were shown slides displaying a succession of still frames of the rotating object, they did not recognize the stills as representing a single object as seen from different points of view. Object recognition required a continuous view of the object as it turned. While infants in these studies were too young to crawl or walk, they would have experienced continuously changing views of objects as they moved their eyes and heads, were carried about, or saw the objects themselves move.[12]

WHEN LIAM WATCHES A BALL IN FLIGHT AND THEN REACHES OUT to catch it, he is momentarily freed from his constant analysis of all things visual. The ball, through its motion, is easy to distinguish from its background, and he can concentrate all his attention on this one easily identifiable target. As mentioned in Chapter 3, our visual system is made up of two systems, a "perception" system for object and place recognition and an "action" system to guide our movements. As I am writing at my desk, I am aware of a coffee cup on my right. I recognize the mug with my perception system. Aware of its presence, I tell myself to reach for it. Yet, having told myself to take a drink, I am not aware, in the moment, of how I am using my action system to move toward and grasp the cup. It all happens automatically. That is a good thing too. If we had to think consciously through every move, we could do nothing else.

The perception and action systems tell us very different things about an object.[13] With our perception system, we are able to recognize an object, such as a chair or coffee cup, at any size or from many different points of view. Indeed, there are neurons in higher areas of the perception pathway that respond to a given object no matter from which direction the object is seen, how big or far away it appears, or where in the visual field it is perceived. Without these neurons, many things would look completely unrecognizable every time we turned our head and viewed them from a new perspective. To the action system, however, the direction from which we see an object is critical.

We must know our position relative to the object in order to reach accurately for it. Perhaps this is why we have two visual systems, one to recognize objects, people, and places and the other to interact with them.

I thought more about the difference between our perception and action systems when Liam wrote about his problem with cup lids. At a coffee shop or cafeteria, he has a hard time figuring out which size lid to pick for his disposable cup. That's a problem for his perception system. Yet, once he has selected a lid, he has no trouble reaching for it and opening his hand to the right extent to grab it and pick it up. Those are problems for his action system. When Liam, Cindy, and I went to Walmart in 2014, Cindy commented that Liam had no concept of length; he did not know the difference between two and six inches. Liam nodded in agreement. Yet, later on that same day, Liam fixed a flat tire on the bicycle that Cindy had brought from her home for me to ride. This required removing the tire with tire irons and replacing the inner tube. Liam did this so quickly and effortlessly that I almost forgot to watch. Even if his perception system confuses lengths and sizes, his action system knew how to deal with the sizes of the tools and parts.

Human infants as young as four months display an impressive ability to reach accurately for moving things. Babies put out their hands to catch a moving object not at the place where it is seen at a given instant but at the place it will be when the object reaches their hand. In other words, they predict the object's trajectory.[14] Indeed, our ability to see and reach for moving objects may be phylogenetically older than our ability to see objects in fine detail. I was reminded of this when I raised my own frogs, our vertebrate cousins. When a fly moves past a frog's eyes, the frog will dart out its tongue to catch it. But if the fly remains absolutely still, the frog will not react. These animals associate food only with moving objects. If I had no live crickets or mealworms to feed my frogs, I fed them some of my dog's food. I couldn't put the dogfood out in a little dish for the frogs to find, however. Instead, I had to attach a clump of wet dogfood to a thread and then bob it in front of the frog's face. Only if the clump was moving would the frog reach for it. For frogs and for people, the ability to see motion in order to move and interact with objects is basic and fundamental.

What's more, moderate physical activity may sharpen our vision. Studies in mice show that locomotion (mice running on a mouse treadmill) enhances the speed and accuracy of their visual cortical neurons in discriminating visual stimuli.[15] Studies in humans suggest that moderate exercise enhances sensitivity in the human visual cortex.[16] Physical exercise also enhances visual plasticity—the ability of visual circuits to change in response to experience.[17] As Liam ran after a thrown ball, he was capitalizing on a great way to enhance his vision.

Motion also plays a critical role for the newly sighted children and young adults from Project Prakash in India. This program, founded by Dr. Pawan Sinha, a professor at the Massachusetts Institute of Technology, provides care for people with curable forms of blindness. Many of the patients were born with dense cataracts over both eyes, eliminating all vision but an awareness of light and dark. Sight can be restored by removing the cataractous lens and implanting an intraocular lens. This surgery is best done within the first months of life; however, many children in India are too poor or live in areas too remote to receive such interventions. Sinha, along with a group of ophthalmologists and optometrists, set up screening camps in rural areas in India, identified children and young adults with treatable forms of blindness, and arranged for their surgeries and care. They took a big risk, however, because many scientists and doctors believed that blindness could be treated only during a critical developmental period that ends by age eight or earlier. After this time window, the brain, it was thought, was no longer flexible enough to handle the onslaught of new visual information. The Prakash patients, many older than eight years, exceeded all expectations. When they first looked upon the world, they, like Liam, saw patches of shapes and colors that did not organize themselves into meaningful objects. However, if an object moved, then all its parts moved as one, allowing the newly sighted patients to link the different parts of the object together and segregate it from the background. Indeed, the first things that the Prakash patients could recognize were objects that either moved on their own or were moved by others, such as animals, cars, and bottles.[18] For Liam, for the patients in Project Prakash, and indeed for all of us, motion may be vision's greatest teacher.

– chapter 7 –

Going with the Flow

When Liam and I bicycled around St. Louis, he rode faster than me. I could only keep up with him because he stopped at each intersection to look carefully in all directions. I, too, am hesitant to make turns while moving, but some of the streets we rode on were so broad, open, and quiet, I could check for oncoming cars and move on without stopping. At one point, we came to a narrow archway that Liam biked through without slowing down, something I couldn't do. He also rode right over curbs, while I stopped and walked my bike. Liam joked that he just didn't see them. Maybe he didn't notice the curbs and thus rode over them without fear, but I think he just has good balance. We entered Forest Park and continued along broad paths until we reached the medical center at Washington University, where Liam was taking classes. As long as the streets and paths were quiet and uncrowded, Liam was comfortable on his bike.

Shortly after his surgeries, Liam discovered that his own motion helped him to interpret the layout of a scene. As he swayed back and forth, objects sorted themselves out in depth. Nearer objects appeared to move faster in his visual field and in the opposite direction to his motion while further objects appeared to move more slowly and in the direction of his motion. This phenomenon, called motion parallax, provides important information about how things are organized in depth. Infants develop a sensitivity to depth from motion parallax by twelve to sixteen weeks of age, and such early development makes ecological sense.[1] Information about object unity and depth layout from motion is much more reliable than, say, information from color or texture. A single object may be made up of many colors and textures but will nonetheless move as a single piece.

We use our vision along with our vestibular system (the balance organs in our inner ears) and our proprioception (sensors in our muscles, tendons, and joints) to keep ourselves upright. Here, vision plays a major role—just compare how long you can stand on one foot with eyes open versus eyes closed. Yet Liam developed excellent balance despite having poor vision throughout childhood. Even before his intraocular lens implants, he loved to snowboard. Having grown up with such poor eyesight, a dramatic improvement in vision could have disrupted rather than improved his balance. But he had adapted quickly.

"Are we going fast enough?" Liam asked at one point. He had a speedometer on his bike and noted that we were going eight miles per hour. This was a comfortable speed for me, but Liam normally biked at ten or, if he was late, fifteen miles per hour. This was impressive because as we cycle (or walk or drive a car), the world around us, though stable, appears to be in constant motion. As we move forward while looking straight ahead, our surroundings on both sides appear to move backward. Turn your eyes and head in a clockwise direction, and the world appears to move counterclockwise. Move your eyes up, and the world appears to move down. This phenomenon, known as optic flow, helps us to solve a visual riddle. When we are stationary but objects move, they cast images that move across our retinas. When we move but the objects are stationary, they also cast images that move across our retinas. So how do we know whether we or the objects are moving? Sensors in our inner ears, joints, and muscles help solve this riddle, but optic flow also plays a major role. As we move, it's not only the images of the objects that move across our retinas. The whole world, objects and background, appears to move of a piece, and this global motion helps us to know that we, not individual objects, are in motion.

Vision psychologist James Gibson recognized the importance of optic flow and gave it its name. During World War II, he was charged with the task of finding out what makes for a good pilot. He discovered that a pilot lands a plane while watching the way the whole world appears to flow by.[2] As the pilot prepares to land, he keeps the runway in the center of his gaze, which causes the land on either side to appear to move forward and expand around him. The landing strip is the focus of this two-sided expansive flow. When you drive a car down the highway, the markers on either side of the lane appear to diverge around you, and the speed of their approach tells you your speed. Moviemakers exploit optic flow all the time. In a car chase scene, we see the action from the point of view of a camera in the speeding car, so the world appears to rush by as though we are in the driver's seat. As the car turns, the optic flow pattern changes, and we feel ourselves heading in a new direction.

Before receiving his IOLs, Liam was confined to a cocoon of vision a few feet in diameter. As he moved then, he didn't see the visual periphery, so he didn't see the world flow by. During recovery from his first IOL surgery, Liam tried to stand up but immediately fell over. Dr. James Hoekel, his optometrist, explained to him that the double-concave glasses he used to wear made everything look smaller. Without the glasses and with his new IOLs, objects looked bigger, as if they were coming toward him. So, naturally, he leaned back and lost his balance. In addition, we move forward when we rise from a chair to a stand, which causes the world to either side to appear to move backward. With Liam's dramatically improved view of the visual periphery, this apparent movement was far more evident and may have caused him to react by leaning backward and losing his balance. That Liam was able to regain his balance quickly and learn to move about with his new vision is a testimony to his adaptability.

As Liam biked along, he kept the center of the sidewalk in the center of his gaze, causing the land on either side to appear to flow or expand around him. As he headed into a curve, the flow expanded more on the outer side. When Liam biked through an archway, its opening appeared to expand the nearer he got, a phenomenon called looming. The faster the opening appeared to enlarge, the faster he was going. To get through the archway, he had to keep that expanding opening in the center of his gaze. All this, Liam learned, probably unconsciously, as he continued to ride day after day.

Optic flow is so important to our sense of our own movement that an area of cortex, called MST (medial superior temporal area), which is part of the "action" pathway, is devoted to processing certain patterns of optic flow.[3] But our sensitivity to optic flow can sometimes cause confusion. Many of us have had the experience of looking through a train window to see our view totally dominated by another train. If the train, seen through the window, starts moving forward, we may momentarily feel like we're moving backward. The same confusion can happen when looking at large clouds sliding rapidly across the sky or while watching a movie on a large IMAX screen. When we see a large portion of our visual field move as one piece, as occurs in the above examples, we interpret the movement as optic flow produced not by objects on the outside but by our own movements.

To read a road sign while cycling, Liam needed to fix his gaze on the sign and then visually track it as he rode by. If he turned his head to the right, he would have to move his eyes rightward to keep the sign in his central vision as he passed it. The closer the sign was to him, the faster its image would move across his retinas and the faster he would have to follow it with his eyes. This task is made easier if the two eyes work together so that both are aimed at the sign at the same time and provide the brain with the same information.[4] I discovered this for myself when I gained better coordination of my crossed eyes through vision therapy. After years of getting lost while driving, I could finally track and read road signs. Poor tracking while moving may help explain why Liam stopped at every intersection to make sure it was safe to cross.

When describing a late-night trip to a grocery store, Liam wrote, "Outside everything was still and I wasn't even thinking about vision. The instant we went inside, everything was moving around me, not spinning or in any direction, just the constant feeling of motion of everything else, not me." These sensations may have been exacerbated by the store's narrow aisles. Items jam-packed along the aisles may have appeared to zoom by, contributing to Liam's unease.

If Liam is talking and must turn from one person to another, the ensuing optic flow can make him dizzy. Indeed, other people who recovered vision in adulthood have described similar problems resulting from optic flow. When Sheila Hocken first gained sight after a cataract operation, she took a walk to the store and found that the ground below her raced by at a dizzying pace. Trees seemed to be coming toward her so fast that they looked like they would knock her down. So, she closed her eyes and let her guide dog lead her the rest of the way.[5]

Most of the time, however, Liam is comfortable with the optic flow generated by his own movements while walking or cycling. But this is not the case while he is riding in a fast car or train. When most of us look out the window of a speeding train, the world does not appear as a blur. Instead, we see an orderly landscape smoothly passing by. How are we able to watch the passing landscape so comfortably when we are going at speeds far faster than our own limbs can take us? For most of our species' history, people never traveled so fast. We are helped by a reflex called optokinetic nystagmus. As we look out a side window of a moving train or car, our eyes reflexively lock onto a given target, perhaps a tree or house, which we continue to follow with our eyes until the target passes out of view. Then our eyes automatically shift back, lock onto, and follow another target. Liam's optokinetic nystagmus response is weak, making a view from a speeding car or train less stable, coherent, and understandable.

Liam was intrigued by a book, *But Now I See*, in which the late Steven Holcomb, a gold medal winner in the 2010 Winter Olympics, relates his own vision story.[6] Holcomb was going blind from a disease of his cornea, yet continued to train as a bobsled driver. This may seem like an impossible feat, but Holcomb guided the speeding sled through turns almost entirely by feel. When a remarkable new treatment restored Holcomb's eyesight, he discovered that he was distracted by his view of the world whizzing by as he raced. His solution was to drive with a dirty helmet so that he didn't see too much.

While walking or bicycling, slower movements under his own power, Liam is able to use optic flow to judge how fast and in what direction he is moving. But that information alone will not get Liam or any of us to where we want to go. We need some kind of mental map that tells us where we are now relative to where we want to be. For most of us, vision, from the use of road signs to prominent landmarks to the position of the sun in the sky, plays a primary role in navigation. How do you find your way, however, if you are just learning to see?

- chapter 8 -

Finding His Way

ONE MORNING, ON A WALK TO THE SUBWAY IN CAMBRIDGE, Massachusetts, I found my way blocked by five businessmen engaged in intense conversation. "Excuse me," I said softly, but the businessmen didn't budge, so I zigzagged my way carefully around them. I don't think they were being rude; they were too immersed in their own discussion to see or hear me.

A short block later, I spied the tip of a white cane on the ground beside me, immediately followed by its owner, a blind man striding past. His cane struck a rise in the sidewalk, and, without breaking stride, he lifted his foot over the obstacle and walked on. He moved so confidently and quickly that I couldn't keep pace with him until he was forced to stop at a busy intersection. "How do you know when it's safe to cross the street?" I asked him as I eyed the walk / don't walk light. He didn't seem in the least surprised to hear a voice off to one side and told me that he listened to the traffic. He described how the cars went first one way, then the other. "I've had years of practice," he said as he started across the street before I even noticed that the walk / don't walk light had changed. The man continued down a sidewalk, had no trouble maneuvering under a scaffold erected over part of the walk, and then turned abruptly into a doorway. This man, though blind, knew exactly where he was and where he was going. Indeed, I realized with some irony, he was far more aware of his environment than the preoccupied businessmen I had encountered earlier.

As Sheila Hocken, who was blind for much of her life, mentions in her book *Emma and I*, people are surprised that blind people know where they are relative to the things around them.[1] To her, this was obvious. "It's your life!" she reminds us. No matter our sensory makeup, we use the information available to us to develop our wayfinding skills. My former student Zohra, whom we will meet next, is deaf and thus very dependent upon her vision. She's an excellent pathfinder. Indeed, on family trips, her family depends on her to find the best routes. While she and I were walking through the streets of Toronto, she described her strategies to me. Pointing to a large building, she explained that she looks for the most salient landmarks, placing them on a map she sees in her head. Most people, she suspects, are not sufficiently selective in the information they take in.

Author John McPhee described heightened spatial ability in a sighted person in his book about former senator Bill Bradley.[2] As a college basketball star, Bradley was known for his perfect shots. If he found no teammate to whom he could pass the ball, then, with his back to the backboard, he tossed the ball over his shoulder right into the basket. When asked how he did this, he said that when you play on a basketball court long enough, "you develop a sense of where you are." Liam, too, even before receiving his intraocular lenses, had a "sense of where you are." Like Bradley, he could shoot basketballs into the basket with his back turned. He had no trouble moving through his house, even in darkness, and moved so confidently that his grandmother did not recognize how little vision he had. With or without vision, we all have to find our way.

In the 1940s, psychologist Edward C. Tolman watched hungry rats find their way to food located at the end of one arm of a maze.[3] After learning the maze, the rats used the same path to reach the place where the food had been in the previous training trials. But they were not dependent upon this particular route. If Tolman blocked off parts of the path, the rats would find alternate routes much more rapidly than in a pure trial-and-error manner. They seemed to have a global sense of their starting point and their goal. They appeared to have constructed a mental map of their environment, and this gave them the flexibility to use many alternate paths. Tolman called this mental representation of space a cognitive map.

Since Tolman's pioneering work, support for the existence of cognitive maps has come from many other studies on animals and people. My favorite involves an incident told by the famous scientist of animal behavior Konrad Lorenz, who raised a gosling he called Martina.[4] When Martina was very young and still flightless, Lorenz would carry her through the nearby village, meadows, and forests or have her walk beside him. One day when she was just learning to fly, he released her into the air at quite a distance from his house. To his horror, he watched her fly out of sight. Lorenz was convinced that she would get lost since she had never flown the route from the release site to home, a route that would take her through a meadow and a forest. He searched all day for her and arrived home at dusk feeling desperate and weary. But there was Martina waiting at his doorstep in great agitation. Lorenz realized that the bird must have created an overall mental map of the local landscape during their previous walks, a necessary skill for an animal that walks, swims, and flies over distances.

The ability to create and use a cognitive map to find one's way varies widely among people. As Liam took me around St. Louis, I learned that, despite or perhaps because of his compromised vision, he has great powers there. He loves treasure hunts and maps of all sorts. This was evident when we went "geocaching" in Forest Park. To find the geocaches, Liam enjoys working with the maps and coordinates loaded onto his portable GPS or smartphone. Sometimes he makes maps of his own—for example, one of the medical library—by first drawing a grid on a piece of paper and then filling in the grid with landmarks. He was surprised and delighted to discover maps of the medical center buildings by the elevators and immediately studied and mastered them. As we walked along interior halls in the medical center during my visit in 2014, I asked Liam about the location of other medical center buildings or the general direction of his apartment. He pointed immediately and confidently to their locations. When I asked him for the compass direction north, he readily indicated the correct direction. (I carried a compass, which confirmed this.) Liam added that he could imagine north on his mental map. But when I asked him if he saw the map in his head, he said, "Not really."

Liam tells me that his visual imagery is very weak; yet poor visual imagery does not preclude his skill at forming and using cognitive maps. Indeed, many people I know who have normal vision and good wayfinding skills tell me that they don't see actual maps in their head but instead have a feeling for where places and things are located. For Liam, his compromised vision has honed his mental mapmaking ability. While a person with excellent vision can look around and quickly discover where things are located, Liam must rely more heavily on his memory and cognitive maps. What's more, he tries to avoid all left-hand road turns when biking or walking, forcing him to use his cognitive map to figure out alternate routes.

Some of Liam's early experiences sharpened his ability to build cognitive maps. When we make a round-trip to and from a particular location, we must imagine our path as heading in one direction when we set out and in the reverse direction when we return. This requires a mirror-image reversal of our map. As a child, Liam learned to write Braille using a slate and stylus. When you punch out the dots with the stylus onto the back of a page, you must place them in a mirror image arrangement to the way you feel them as you read. Liam learned this task easily, a skill that later made it easy for him to understand mirror-image (stereo) isomers in organic chemistry class, a concept some students struggle with. So Liam was proficient at imaging spatial configurations and rotating them about in his head, and these skills help with the construction and use of cognitive maps.[5]

The hippocampus and surrounding cortex are probably the areas in the brain where cognitive maps are created, stored, and recalled in people and other vertebrate animals.[6] Since the hippocampus receives input from several modalities, including vision, hearing, touch, and self-motion, all these inputs contribute to the creation of cognitive maps. For those of us with sight, vision plays a primary role in map building, but, as Liam taught me, a good mental map can be created without vision. Creating a cognitive map of his environment was not a problem for Liam. Judging distance with his vision, however, was an entirely different matter.

DURING MY VISIT TO ST. LOUIS IN AUGUST 2012, LIAM TOOK ME up to the top of a hill in Forest Park. As we paused to take in the wide view, Liam pointed out that he's not sure, when seeing two parallel paths in the distance, whether they're a few steps away from each other or much further apart. He usually thinks things are closer to him than they actually are. So, for determining distances, he uses his smartphone or portable GPS. This reminded me of how important it was that Liam had sufficient vision as a young child to learn to read print and how smart he was to revive and refine this skill once he received the IOLs. He can read 10-point font, so his smartphone is of great use for navigation. And he was smart too to keep up his Braille reading skills, for on bright, sunny days, it's hard for him to read a smartphone screen outdoors. Ever resourceful, he'll connect his Braille display, which looks like a small keyboard, to his smartphone to get the information he needs. The Braille display also comes in handy for reading books on a bus.

Liam's trouble with judging the distance between two paths is part of a general difficulty in using his vision to interpret the three-dimensional spatial layout. It is one thing to recognize individual objects but another thing entirely to understand the organization of a whole landscape. From the top of that Forest Park hill, he struggled to understand the distance between things or their actual size. As described below, he did not use perspective and shadows to help sort things out in depth and space. Such cues are used by infants starting at about seven months of age, and they come automatically to most of us.[7] Liam, by contrast, had to consciously create his own strategies to understand depth and spatial layout. As we continued to explore the park and then the Washington University Medical Campus, Liam explained his strategies to me, and I tried to imagine the world as he saw it.

WE USE OUR VISION TO SEE OVER AN ENORMOUS RANGE OF DIStances. Scrutinizing a close target, we can see, as separate dots, two specks that are as close as one-fifth of a millimeter apart. Looking up at the night sky, we can see stars that are lightyears away. Without special equipment and calculations, however, we have no way to judge the distance of those stars from each other or from ourselves. For that matter, the moon is very much closer to us and smaller than the sun; yet we cannot make these judgments with vision alone, except during a solar eclipse. But we can use our vision to judge the distance of Earth-based structures. As vision psychologist James Gibson pointed out, we can use the ground as a reference.

Gibson realized the importance of the ground for distance judgment after he was assigned by the military to study student pilots during World War II.[8] Everyone assumed that people with the best depth perception would make the best pilots, but scores on standard depth-perception tests did not predict piloting ability. When landing an airplane, Gibson concluded, a pilot didn't judge distance by looking through empty air. Instead the pilot used his sense of the spatial layout of the ground. Perhaps we had to take to the air to realize just how important the ground is for understanding how we organize our visual world.

In empty air, it's hard to tell how far or how big an object is. A few years ago, while out walking, I spied what I thought was a bird flying by some distant trees. An instant later, I realized that the flying creature was much closer to me, and it instantly decreased from bird to insect size. I was watching a butterfly. These changes in perceived distance and size were not conscious. They happened together and spontaneously. I was experiencing the phenomenon of size constancy. As objects move toward us, they cast larger images, and as they move away, they cast smaller images on our retinas. Yet, for a considerable range of distances, we see the images not as the size they project on our retinas but as life-size.

As an example of size constancy, think of a basketball (30 inches or 75 centimeters in diameter) as it appears to you when you hold it at the end of your arm. At this close range, the basketball takes up a lot of your field of view, a field that spans 155 degrees of visual angle from the far left to the far right (with your eyes at the angle's vertex) and 135 degrees from the bottom to the top. Indeed, the basketball covers a visual angle of about 64 degrees. If instead you see the ball halfway down the basketball court (47 feet or 14.5 meters away), the balls covers a visual angle of only 3 degrees, and the ball located a full court (94 feet or 29 meters away) takes up a visual angle of a mere 1.5 degrees. From holding the ball on your arm to seeing it a full court away, the ball's visual angle decreases by 97 percent; yet it still appears life-size, an apparent contradiction! Of course, there is a limit to this size constancy. The ball coming directly at your eyes will loom large at very close range, and people on the ground look like ants when seen from the top of a skyscraper. Size constancy is also poorer in the vertical dimension—that is, when looking up. Many of us have been surprised by how large one of those green signs posted over the highway actually is if we see it on the ground. The same is true for stoplights. Nevertheless, for an impressive range of distances, we can judge how far away and how large an object is, and this judgment depends upon our view of the spatial layout.

Like me with the bird-turned-butterfly, we all need references to judge size and distance. And the ground under our feet is one very informative and (barring earthquakes) stable reference. On flat ground, a person standing six feet (1.8 meters) tall can see for about three miles (4.8 kilometers). Since the distant horizon appears at eye level, all objects closer than the horizon that are resting on the ground appear lower down in our visual field. So the height of objects in our field of view gives us information about their distance from us—the higher up, the further away.

When I first met Liam in 2010, five years after his surgeries, he pointed to his backpack, which lay on the floor several feet from where he was sitting. He said that the carpet beyond the backpack appeared higher than the backpack itself. Indeed, the carpet was further away so appeared higher in his visual field, but, to Liam, the backpack then looked to be below the carpet and therefore appeared underground! Four years later in 2014, when Liam and I were walking together, we came upon several rows of tables outside a café. While he knew that the more distant tables were behind the ones in front, the distant tables still looked to him as if they were standing on top of the nearer ones. Most of us instantly and unconsciously interpret objects higher up in our visual field as further away. Liam, however, had to think through this interpretation.

Our understanding of linear perspective comes into play here too. As we look into the distance, parallel lines, like the sides of railroad

FIGURE 8.1. To Liam, the girl, whom he saw down the hallway, looked like she was standing on a trapezoid.

tracks or highway lane markers, appear to converge in the distance. Shortly after Liam received his IOLs, he spied a girl at the end of a hallway. As the sides of the hallway appeared to narrow, they looked to Liam like they were going up, not out. So the girl seemed to be standing not further away but on top of a trapezoid instead.

It's not just Liam's lack of visual experience that makes distance judgments less automatic; poor stereovision plays a role. Although I was stereoblind until age forty-eight, I always thought that I saw a straight road as extending out beyond me in a flat plane perpendicular to my face. When I gained stereovision, I was surprised by how much more a road appeared to extend along the ground. It appeared much more outward and much less upward. Thinking back to my old way of seeing, I could understand how Liam saw the tables as arranged as a vertical stack even though he knew they were located one behind the other.

Closely related to linear perspective are texture gradients, and these too were first studied by Gibson.[9] As we look across a tile floor, the tiles further away appear smaller and denser than those close up. The same is true for grass blades in a meadow or evenly spaced telephone poles along the road. When we combine our understanding of texture gradients with a sense of distance gained through linear perspective, we have a powerful way to judge distance and size. This judgment is immediate and automatic and can even be used with optical illusions to fool us.

In the Ponzo illusion in Figure 8.2, we see a set of railroad tracks receding into the distance. The lower gray horizontal line appears less wide than the gray horizontal line higher up on the page. Actually, the two gray lines are the same width, but we see the upper one as wider because, with the convergence of the lines of the railroad tracks (linear perspective) and the increasing density of the horizontal track lines (texture gradient), we interpret the upper gray line as further away. If the two lines are the same width, as in the picture, but we interpret the upper one as further away, then in real life it would actually be longer. That is how we see the illusion. This interpretation is so

FIGURE 8.2. The Ponzo illusion. Are the gray horizontal lines the same length?

automatic that it is almost impossible to see the two lines as being the same width even when we know that they are.

When, in 2014, I showed Liam the Ponzo railroad track illusion, he saw the railroad tracks as a design going up on the page, not as rails receding into the distance. As a result, he wasn't fooled by the illusion and saw the two gray lines as being the same size. It's possible that he has rarely seen railroad tracks or pictures of them, so he did not recognize the image. But I think the more likely explanation is the one Liam himself has given. A childhood of near blindness robbed him of the experience of seeing things spread out in depth. So he tends to interpret real-life scenes and pictures of real-life scenes as two-dimensional. His interpretation of the tracks

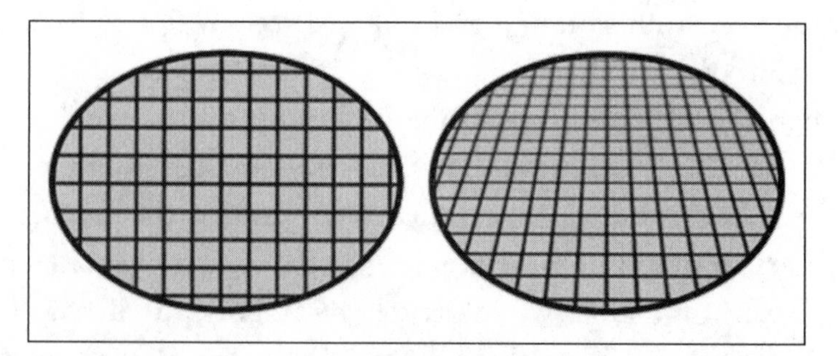

FIGURE 8.3. The surface inside the right oval appears to recede in depth due to our interpretation of converging lines and texture gradients.

in the illusion is similar to his view, mentioned above, of the girl on a trapezoid or his view of the tables in the café as stacked one on top of another.[10]

In Figure 8.3, the intersecting lines in the left oval form a flat, vertical grid. But the converging lines and texture gradient inside the right oval create a surface that appears to recede in depth. However, Liam saw the surface inside both disks as flat. When observing the Ponzo illusion and the disks, he did not use linear perspective and texture gradients to judge distance and size.

So I was really surprised and impressed by Liam's interpretation of the corridor illusion (Figure 8.4). This picture looks like a view of a tiled corridor with two black sticks or poles that are actually the same height. Given the perspective provided by the converging lines and the texture gradient provided by the four-sided tiles, however, we see the pole on the right as further away and taller than the one on the left. When I asked Liam what he saw, he replied, "The sticks slowly changed between the same size and different sizes, and most of the time I saw both at once, with the front stick (with fewer wall lines intersecting it) being slightly smaller." Liam's interpretation was unstable. The sticks looked to be both the same and different sizes

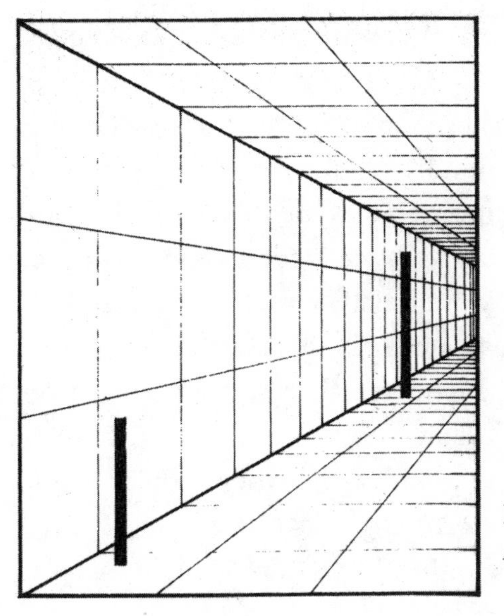

FIGURE 8.4. The corridor illusion. Are the poles the same height?

at the same time. In his typical analytical fashion, he also concluded that the more distant stick should look taller because it intersected more wall lines. In other words, Liam used texture gradients in a very conscious way to analyze the picture. Since, with his initial view, he spontaneously saw the sticks as changing in size, he may have started to use texture gradients in an unconscious way to interpret the height of the sticks. The more he can interpret the scene in an automatic and less analytical way, the faster and easier it will be for him to understand the landscape.

JUDGMENTS OF HEIGHT, DEPTH, AND SIZE COME INTO PLAY EVERY time we climb up or down a set of stairs, and stairs are still a serious challenge for Liam. After Liam's vision improved with his IOLs, Cindy would turn on a light when she saw him approach a staircase, but he would object. Negotiating stairs was easier for him before his vision improved. Climbing up and down stairs using his sight required intense analysis. Liam wrote, "The upstairs are large alternating bars of light and dark and the downstairs are a series of small lines. My main focus is to balance and step IN BETWEEN lines, never on one . . . Of course going downstairs you step in between every line but upstairs you skip every other bar. All the while, when I move, the stairs are skewing and changing, and if you move around a staircase extremely, some weird stuff can happen, like upstairs can have the appearance of downstairs."

Liam's description of stairs sounded totally foreign to me. I had no idea what Liam meant by skipping every other bar while going upstairs and stepping between every line while going down. But then I took a look at a staircase as if it were the 2D pattern that Liam sees, and I saw what I had never noticed before: the bars and lines.

Indeed, when Liam asked his friends for their strategies in recognizing and negotiating stairs, they didn't know what to say. Climbing stairs was so commonplace and automatic that they didn't have to think about it. We may use the alternating bars of dark and light to help us distinguish one step from another, but we don't see the steps as

FIGURE 8.5. Note how the stairs create patterns of bars and lines.

bars or lines in just two dimensions. Instead, the alternating light/dark pattern provides information about both height and depth, indicating both the rise and flat part of a step. For example, the bars or lines appear closer together for the steps at the far end of the staircase, and this texture gradient tells us about distance and height. But Liam did not mention the increasing closeness of the lines to judge the overall height of the staircase.

What's more, stereovision helps in seeing the steps as three-dimensional. Liam's stereovision is poor, so his view of the whole staircase was considerably flattened, appearing perhaps in one plane in front of him. Sacks, too, mentioned that steps were particularly hazardous for Virgil because he saw them as "a flat surface of parallel and crisscrossing lines; he could not see them (although he knew them) as solid objects going up or coming down in three-dimensional space."[11] I know what this is like because I was cross-eyed and stereoblind until age forty-eight. When I was a student, I often encountered a staircase in the university's medical library in which the grout and tiles from one step were exactly aligned with the steps above and below. Many people probably thought that the tiling job was excellent, but for me it was a hazard. Without stereovision, it was hard to tell where one step ended and the next began. So, if I had a lot of books to carry, I would take several trips. I also avoided outside steps at high noon when the shadows were very short, which made it hard to use shading to help gauge the height and width of the stairs.

Liam avoided staircases that were not enclosed by walls on both sides, and he did not like open balconies. When we walked inside large rooms in the medical center during my 2014 trip, he steered us away from the fancy, open staircases in the middle of large foyers (although he could climb them if necessary) and led us instead to the enclosed stairs on the sides of the buildings. These stairs were hemmed in by walls on both sides and were often dirtier and not meant for the public, but Liam preferred them.

It was not just stairs; any surface with uneven footing created challenges for Liam. He took his first airplane ride in 2017. Knowing that the footing would be tricky when stepping on and off the plane, he chose not to trust his vision and used his cane instead. Liam avoided the St. Louis MetroLink (the city's light-rail system) for years until a blind friend showed him how to negotiate the train platform with a cane.

LIAM'S DISTASTE FOR OPEN STAIRWAYS EXTENDED TO ANY SITUAtion involving an abrupt change in view. He didn't like mirrors or windows, both of which frame a three-dimensional scene that is not continuous or connected with the area right around it. This is how Liam put it in 2012: "I still don't have anything to do with mirrors . . . They disrupt my world too much for me to deal with right now. This reminds me of one of my very basic vision rules . . . The rule is Mirrors—same side; Windows—other side, because to me they are both just complex visual information inside a box or frame." It took me a moment to understand what Liam meant by "same side" and "other side." He was referring to inside and out. An indoor mirror reflects a scene from the inside of a room (unless its reflection includes what is seen through a window), while a window, as seen from indoors, provides a view of the outside.

Imagine standing inside a room and looking through a glass window. You see a view of the outdoors, but in many instances you also see reflections of images that are inside. This can make for a very confusing picture, especially for Liam, as the photograph in Figure 8.6 shows.

FIGURE 8.6. Reflections through a plate-glass window.

IN ADDITION TO PERSPECTIVE AND TEXTURE GRADIENTS, WE UNconsciously use other cues, such as shading and shadows, to gauge depth. For example, in Figure 8.7, most of us see bumps and pits. After a moment of careful inspection, you might notice that the bumps are light on the top and dark below, while the pits show the reverse pattern. Our experience has taught us that light almost always comes from above. The light will reach and brighten the top surface of a bump but will be blocked from the bump's surface below. The reverse lighting pattern happens with a pit. If you turn this page upside down, the bumps become pits and vice versa.[12] For Liam, with far less visual experience, the bumps and pits in the above image used to look flat. Now, they are looking more like the bumps and pits everyone else sees, and it's getting harder to switch back to his old interpretation

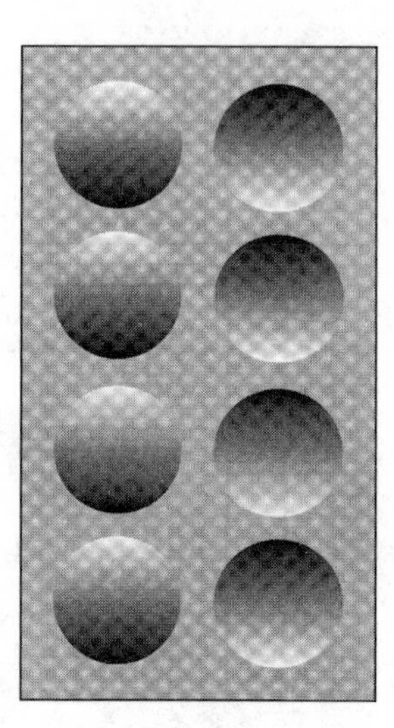

FIGURE 8.7. Shading creates a sense of depth. Turn this figure upside down, and the bumps become pits and vice versa.

and see them as flat. His ability to use shading to interpret solidity and depth is becoming more automatic.

However, Liam has a hard time with shadows. While walking on a sidewalk, he is not sure whether a line shadow on the ground is indeed a shadow that can be ignored or a stick that could trip him. "There are things," Liam wrote, "that light can do that can add lines to any surface or multiple surfaces (which I affectionately call 'lies'), and I have to figure out what lines to disregard and which are pertinent in addition to deciphering what they mean." So, on a shadow-filled day or when walking with others, when he is expected to talk at the same time, he uses his cane to test for obstacles and shadows.

The shadow of a pole on the sidewalk emerges from and is continuous with the bottom of the pole. But some shadows are cast some distance from their source, and they give us additional clues about an object's location. In the upper part of Figure 8.8, the shadows are attached to the balls, while in the lower part they are cast further away. The detached shadows cause the balls to appear to float above the floor. But the balls in the two parts of the figure all looked to be in the same place to Liam; the balls in the lower figure didn't seem to be floating above the floor. To him, the lower three balls and shadows

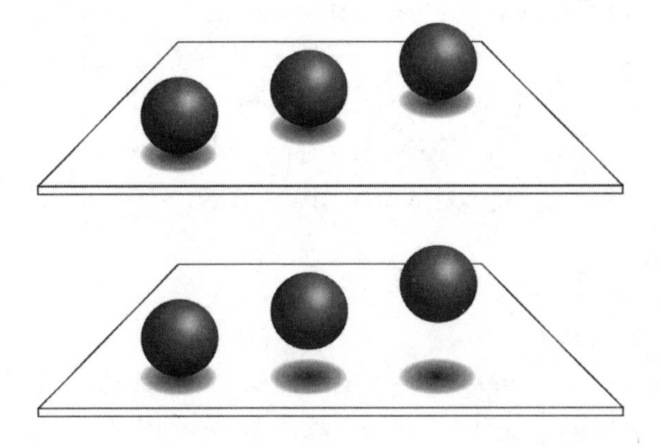

FIGURE 8.8. Shadows affect our interpretation of the balls' location.

made a sideways *V* of five spots. He did not use the shadows to interpret the balls' positions in three-dimensional space.

Liam's difficulties in interpreting landscapes affected his ability to appreciate pictures. When a newly sighted person first sees a picture, he may not recognize it as a picture at all.[13] He cannot discover the contents of the picture by touching it, and the concept that the picture represents a miniature view of a real scene may seem totally foreign to him. Indeed, he may attend less to the picture and more to the surrounding frame. Here again, Liam's vision in childhood, albeit poor, offered him an advantage. Liam was exposed to pictures as a child, so he recognizes them as pictures and appreciates simple drawings and cartoons. When I showed him the visual riddle, or "droodle," in Figure 8.9, with its clever caption, he immediately got the joke.

FIGURE 8.9. "An early bird who caught a very strong worm."

However, Liam doesn't like more complicated pictures unless they are abstract. Pictures of landscapes, which make great use of perspective, texture gradients, shading, and shadows, he likes least of all. This is in great contrast to his love of maps, which represent spatial relationships, such as distance and direction, but do not necessarily include images.

SINCE SHADOWS AND REFLECTIONS OFTEN CONFUSE LIAM, HE relies more heavily on motion-based cues to recognize objects and interpret landscapes. He may identify an object by walking around it to see it from multiple viewpoints, and swaying back and forth helps, through motion parallax, to sort things out in depth. From the start, we all, like Liam, prioritize motion-based information. Infants are sensitive to motion-based cues at four months but do not develop an ability to exploit shading and shadow until around seven months of age.[14]

Yet, even motion-based information can confuse, as Liam demonstrated to me when, during my 2014 visit, he took me up to the fancy glass walkways that linked the medical buildings at Washington University (Figure 8.10). Since the glass passageways were at least one story high, our view of the passing landscape was different from that seen from the ground. Liam saw the cars and people below not as life-size but as very small instead (Figure 8.11).

FIGURE 8.10. Glass walkway.

FIGURE 8.11. View from glass walkway.

This is a problem with size constancy. What's more, stationary objects outside that were closer to the window appeared to move more quickly as he walked than those further away. This effect of motion parallax wasn't helpful, as it would have been on the ground, and threw him off balance. So Liam took out his white cane and pushed it along the passageway, always keeping it in contact with the ground. He kept his eyes on the carpet pattern so he could walk right in the middle of the passageway, as far from the windows as possible. Liam felt the same discomfort while riding on the MetroLink, the St. Louis subway, when the train moved onto an elevated track. For Liam, understanding the landscape took on a whole new challenge when seen from one story above.

Liam's trouble in gauging three-dimensional layout and the distances between things initially made crossing streets very hazardous. When he came to an intersection, he wasn't sure how far away the approaching cars were. If he encountered an intersection with a road directly in front of him, he couldn't tell if the cars were approaching the intersection or were already there. So he avoided busy intersections, particularly those without crosswalks and stoplights, and went out of his way to avoid left-hand turns. Crossing streets could set his heart racing. When this happened, he took off running after crossing the street because it felt better to have his heart pounding from intense exercise than from panic.

After his surgeries, Liam wanted to redefine himself as a visual person, and he put great pressure on himself to meet this expectation. Renouncing the cane, he attempted to navigate the world primarily with his vision. When, in 2016, he worked as a medical scribe at a hospital located eight miles from his apartment, he had to walk or bike to a bus stop two and a half miles away. In total, the trip to work was a one- to two-hour commute over varying landscapes that flowed by as he walked along the ground, pedaled by on a bicycle, or rode up above the ground on a bus. He managed well but only when he worked at night and was not bothered by the glare of the sun or confusing shadows.

"Focusing so hard on sorting out what is in front of me takes an athletic level of focus," Liam told me. And all that attention directed to what is front and center means he doesn't take in the overall landscape. Once, while walking along a sidewalk, he saw a sign indicating that he needed to detour around a construction site. The sign pointed to the construction site, but he didn't see the construction, only the pointing arrow. So he followed the direction of the arrow and would have collided with a large dump truck if a construction worker hadn't interfered. Liam was horrified. He was seeing but not taking in the wider view.

Indeed, Liam is in a unique situation, even among people with vision loss. In the doctor's office, under even lighting and while sitting still, he may look at an unmoving chart of high-contrast letters and be able to read the 20/60 line. This acuity level is considered good enough to move about without special nonvisual accommodations. Yet, outside he may be blinded by bright sunlight and glare. Moreover, his performance with the eye chart doesn't uncover the visual-processing deficits he struggles with—deficits that are so hard for the rest of us to imagine. Liam has friends who are legally blind; their visual acuity is worse than 20/200, but they understand and negotiate the landscape with their vision better than he can. His friends may not be able to see fine detail, but they may have had normal vision in their early years, allowing them to instantly and effortlessly understand the big picture.

But Liam has one advantage that many of his friends, who suffered late vision loss, don't have. Since he was practically blind since childhood, he is experienced in navigating with a white cane. Over time, Liam discovered that he could travel best by using a combination of his vision, his cane, and the GPS. The cane lets him know what is immediately in front of him, freeing him to look further into the distance and take in the overall landscape. Using his unique combination of skills, Liam has found a way to travel with confidence, opening up the world to him and allowing him to explore new places.

– chapter 9 –

Christmas Lights on the Grass

FIGURE 9.1.
Liam, 2019.

LIAM HOPES TO HELP OTHERS WITH VISION LOSS, SO HE HAS immersed himself in the biomedical field. He wants to learn as much as he can about the eye and vision and reads ophthalmology textbooks in his spare time. He has worked at a hospital emergency room and ophthalmology practice as a medical scribe and also in Dr. Tychsen's primate-vision research laboratory. His lab job was to assemble photo montages from neuroanatomic slides of the visual cortices of strabismic monkeys. Infant monkeys with strabismus are the best animal model for revealing what goes wrong in the visual cortex of a strabismic child. Creating each montage was like piecing together a puzzle. Each slide showed a small region of the monkey's visual brain, but when all the slides were correctly assembled, the complete montage revealed the whole unified and reconstructed visual cortex.

Liam loves puzzles, so he embraced the task of assembling the montages, but the work presented new visual challenges. To make the montages, he first had to view the slides under a microscope and then memorize what was on them. But, as soon as he put a different slide on the microscope, he forgot the first. As always, he did not shy away from the problem but broke it down into parts and solved the parts, one by one. First, Liam made a sketch of each slide and, using these drawings as a reference, then selected the best slides for the montage. With practice, he could sketch one slide, turn back to the photo gallery, which showed thumbnails of all the slides, and know which slide should be viewed next. When constructing the montage, he made a note of a blood vessel or stained cell on the right edge of one slide and then matched it with the same structure on the left edge of another. To do so, he had to move his eyes smoothly from one slide to the next. Remembering the images on the slides and tracking with his eyes, Liam told me, was initially difficult but got easier with practice. He was able to recognize when his eyes jumped instead of moving smoothly. Liam's tedious work paid rich dividends; he was able to reconstruct beautifully the whole visual cortex of three monkeys. This work involved the assembly of hundreds of slides, a task even someone with normal vision would find daunting.

For Liam, every new experience brought visual challenges. He worked so hard on his new vision that Cindy sometimes wondered whether he should have undergone the IOL surgeries at all. She learned that some who gained sight after childhood became depressed, even ill, and rejected their vision. She had hoped that the doctors could tell her what Liam would be able to see, but they were learning from Liam too.

As for Liam, he is quite definite about the results of his operations and his new sight. He has developed his vision and restructured his perceptual world for what he enjoys most—the ability to navigate on his own, undertake interesting jobs in the medical and scientific fields, and engage in games of all sorts. Using his vision together with his cane and the GPS, Liam now seeks out and enjoys trips to new places. He has overcome much of his shyness, plays on sports teams for the visually impaired, and has joined advocacy groups for people with low vision and albinism. Through these activities, he has found new friends. "I can't imagine the past few years any other way," he wrote to me.

Liam draws on the support not only of a wise and loving mother but of a doctor who has known and cared for him since he was two years old. He rarely shows strong emotions, but at the end of one of my visits, as we waited at an intersection to cross a busy street, Liam grew quiet and then spoke softly but with deep feeling. "My eyes are also Dr. T's eyes," he said, "because he's cared for them for twenty years. I take care of my eyes and see as well as I can because I owe it to him. I'm very thankful."

For Cindy, despite all the difficulties, there were times when watching Liam learn to see had all the magic of watching a very young child first discover the world. Since vision for Liam is such hard work, he rarely describes what he sees as beautiful. But Cindy remembers one morning when she and Liam were up with the sun and Liam saw dew for the first time. "It's like Christmas lights on the grass," Liam said.

PART II

ZOHRA

A handicapped child represents a qualitatively different, unique type of development . . . If a blind or deaf child achieves the same level of development as a normal child, then the child with a defect achieves this in *another way, by another course, by other means.*

—Lev Vygotsky, *The Collected Works of L. S. Vygotsky, vol. 2: The Fundamentals of Defectology*, ed. Robert W. Rieber and Aaron S. Carton (New York: Springer Nature, 1993), (emphasis in the original).

– chapter 10 –

Everything Has a Name

As I stood before my students on the first day of class in January 2010, I was unnerved by a young woman, sitting front row center. She kept her eyes on me, following my every move. Did I look funny? Quickly, I looked down at my clothes but saw nothing amiss. A moment later, I relaxed. My vigilant student was Zohra Damji. She had stopped by my office a few days earlier to explain that she was profoundly deaf and wore a cochlear implant. No wonder she looked so focused; she was following my lecture with a combination of listening and lip-reading.

About one week later, Zohra stopped by my office to turn in a class assignment. As she was about to leave the room, I called after her. When she turned around, I asked her if she would be willing to tell me her story. She gave me a shy but confident smile, as if to say, "No one has ever asked."

The blind and deaf educator and activist Helen Keller once said, "Blindness cuts us off from things, but deafness cuts us off from people."[1] While both sight and hearing allow us to sense things beyond arm's reach, sight provides our primary means for localizing ourselves in space and moving within it. Although we both see and hear a person when they talk, hearing (outside the Deaf signing community) is our main sense for communicating with others. Liam developed his vision primarily to function better in the physical world. But Zohra learned to hear, in large part, to connect with the social world.

Zohra was born in Moshi, Tanzania, a town situated on the lower slopes of Mount Kilimanjaro, Africa's tallest mountain. At that time, her parents lived and worked 274 miles away in Dar es Salaam, but her mother always came back to her hometown of Moshi to give birth to her children. Zohra was the third child. A fourth child would be born three years later.

Shortly after her birth, Zohra's parents took her back to Dar es Salaam, but after six or seven months, they knew something was terribly wrong. Zohra could not lift her head; she slept all the time. There was a good physical therapy program at Kilimanjaro Christian Medical Center in Moshi, where Zohra's maternal grandparents and her aunt Najma lived. So her parents decided to leave Zohra with Najma, who could take her to therapy. To their great relief, the therapy worked, but that was not the end of their worries. Zohra didn't respond to toys that made noise. Could she have problems with her hearing? A friend in Moshi directed Najma to an acquaintance in Nairobi who was familiar with deafness and who recommended that Zohra be examined at Portland Hospital in London. So the family agreed that Najma should take Zohra to London.

Although Najma is fluent in English (and Swahili), her mother tongue is Gujarati, a language from a state in western India. Her grandparents emigrated from India initially to Zanzibar, an island off the coast of Tanzania. Over the next two generations, the family settled in East Africa, but they kept their Gujarati traditions. Najma is an observant Muslim. She had never been overseas, and London in the late 1980s was not as cosmopolitan as it is today. When she visited London with infant Zohra, she was keenly aware that she, with her head scarf, stood out on the London streets and subways (Figure 10.1).

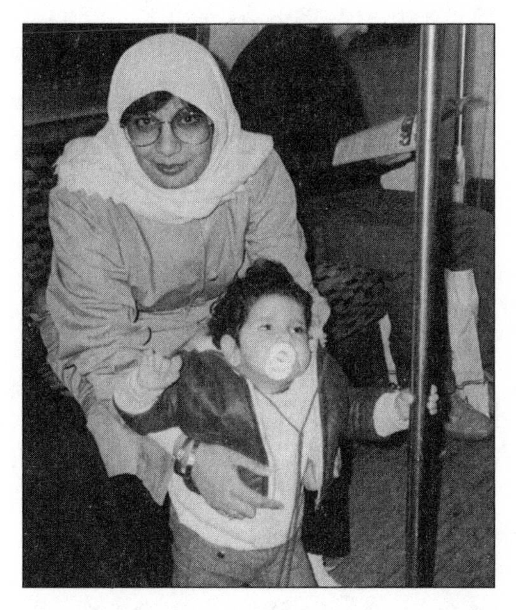

FIGURE 10.1. Najma with Zohra on the London Underground.

The London doctors determined that Zohra could hear no sounds softer than ninety decibels. Although she may have been able to hear a noise as loud as a gas lawnmower blasting into her ears, her hearing was far too poor to pick up the words of normal speech. Yet, in one way, Zohra was lucky. Her deafness was diagnosed very early. At that time, young children were sometimes not identified as deaf until they were several years old.

Zohra was fortunate in another way. A deaf child who doesn't grow up in the Deaf signing community may feel isolated in the hearing world. But Zohra had Najma. They were always together and in constant communication. Their strong and enduring relationship was the foundation on which Zohra developed a strong connection with other people and the world. When Zohra received her cochlear implant twelve years later, she could use it to maintain and enhance those connections.

Najma stayed in London for three months to work with a speech therapist. Zohra was given hearing aids, and the first task was to encourage her to use what hearing her aids provided. So Najma would set out some blocks and, in a loud voice, say, "Go!" If Zohra heard the sound, she moved a block. Then Najma rewarded her and repeated the game, only moving further away or saying "go" more softly. No one mentioned Sign, the rich visual languages of the deaf, conveyed through movements rather than sound. Najma knew no one who communicated with Sign or was part of the Deaf community. So Zohra was trained with the "oral method."

The doctors told Najma that it was up to her to teach Zohra to talk. So when Najma and Zohra returned to Moshi, the family agreed that it was best to leave Zohra with Najma, who could devote herself full-time to the child. Najma worked tirelessly and kept a diary of Zohra's progress. Initially she spoke to Zohra in English and Gujarati but soon switched exclusively to English. It would be simpler, Najma reasoned, to learn just one language, and English was spoken by more people around the world. When Zohra was one and a half, she could recognize a few spoken words such as "ama" (mama), "nanima" (Gujarati for maternal grandmother), and "bye-bye," and she spoke her first words too: "ama" and "up." These first words included the names for the people most important to Zohra. She called both her aunt Najma and her mother "ama" and thought of them both as her mothers.

With each new word, Zohra, who had been so lethargic and passive, grew more active and alert. Soon she recognized words for the parts of the body (hand, finger, foot, eye), household items (clock, soap, towel), food (apple, banana, egg), animals (butterfly, elephant, fish), some actions (pour, open), and some simple phrases too ("close your eyes," "give me"). But it was harder for Zohra to make her own sounds, though "nanima," the word for grandmothers in general and her grandmother in particular, she said distinctly.

To us today, it seems evident that language is learned. We spontaneously speak to infants in "motherese"—a slow, expressive singsong with simple and repetitive words and phrases. With these sounds, children learn the meaning of words and eventually begin to talk themselves. Children who are deaf will obviously have great difficulty developing a spoken language in this way (though they will readily learn Sign). Yet David Wright describes in his memoir *Deafness* that the failure of the deaf to speak was initially blamed not on an inability to hear but on mental deficiency.[2]

For centuries, the deaf were considered unteachable and ineducable. This rigid mind-set began to soften only in the 1500s when Girolamo Cardano, an Italian polymath, read a book titled *De inventione dialectica* by Rodolphus Agricola of Groningen. In it, Agricola described a man born deaf who learned to read and write. Inspired by this account, Cardano proposed a concept considered radical at the time—that language could be separated from the production of sounds. He suggested that a deaf person could "hear by reading" and "speak by writing." (He also proposed that the blind could be taught to read and write through the sense of touch.) While there are no records that Cardano put his ideas into practice, the Benedictine monk Pedro Ponce de León did. In the late 1500s, he taught deaf mutes to write and speak. His motive was religious, since a person must speak to make confession. But most of his pupils were sons of Spanish aristocrats, and these landowners were driven by a more material motive. They were anxious to teach their deaf sons to speak because only those who could speak could inherit property.

Starting in February 1990, Najma took Zohra to Nairobi for further instruction. Since this trip involved a six-hour bus and car ride and the crossing of the border into Kenya, they stayed for three weeks at a time. Their therapist, Mrs. Elizabeth Couldrey, didn't give up on any of her students. Her teaching methods echoed Cardano's idea that a deaf person could "hear by reading." She instructed Najma to make several scrapbooks containing pictures of things in Zohra's everyday world and the accompanying words for them. In one scrapbook, for example, there was a photo of a shoe with the word "shoe" written beside it. Najma would point to the photo and the word, say "shoe," and then have Zohra repeat the word. Najma combed through hundreds of magazines, looking for pictures. With all this training, Zohra learned that everything has a name.

We take our use of names for granted, as if most of the objects out there in the real world naturally come with names. But names are our invention. Even the concept of names must be learned. When Najma described her scrapbooks to me, I was reminded of the blind and deaf Helen Keller and her epiphany at the water pump.[3] Helen was almost seven years old and without language when Anne Sullivan first came to teach her. Using her fingers, Sullivan spelled words into Helen's hands, and Helen soon learned to associate the finger movements for "cake" with the tasty treat, though she did not yet realize that the fingers spelled out a specific name. But a few days later, at the water pump, Sullivan spelled "water" into one of Helen's hands as the water cascaded over the other, and a life-changing transformation occurred. Helen, in *The Story of My Life*, writes, "Somehow the mystery of language was revealed to me. I knew then that 'w-a-t-e-r' meant the wonderful cool something that was flowing over my hand. That living word awakened my soul, gave it light, hope, joy, set it free! . . . Everything had a name, and each name gave birth to a new thought. As we returned to the house every object which I touched seemed to quiver with life. That was because I saw everything with the strange, new sight that had come to me." Helen learned thirty new words that day.

Helen Keller's experience is not unique. In an extraordinary book, *A Man Without Words*, Susan Schaller describes the way she taught Sign to Ildefonso, a twenty-seven-year-old language-less deaf man.[4] The major breakthrough came when Ildefonso finally understood that Schaller's signs represented names for things; the first sign and name he learned was "cat." When this revelation hit, he sat up, rigid, then slowly looked around the room, taking in all the objects in a new way. He slapped the table demanding to know its sign, pointed around the room, asking for the signs for book, door, clock, and chair. Then, overwhelmed by this new knowledge, he broke down and sobbed. Michael Chorost, born severely deaf, learned English through pictures but not until his deafness was diagnosed at age three and a half.[5] As he described in *Rebuilt*, acquiring words transformed him from a "mute, fearful little animal" into a spirited boy. It was as if a lightbulb went off, his mother remembered.

Helen Keller, Ildefonso, and Michael Chorost were old enough when they finally began to acquire language that they could remember learning their first words, and their accounts reveal how much language awakens our minds. Language both reflects and extends our thinking and perception. A word is a symbol that, depending on context, may represent something very concrete or something much broader.[6] And recognizing that an item is part of a basic category is an important part of perception. When Liam learned to recognize things with his sight, he recognized them—a tree, a dog, a staircase—as part of a category. When my son Andy was just learning to speak, he would lie down with his head on the floor and roll a toy car back and forth while saying the word "Volvo." We had a Volvo car at the time and usually referred to it by its brand name. Did Andy think that Volvo was the name for all cars or even for all things with wheels? When did the name represent to him only one brand of cars? The category of things represented by the word "Volvo" changed as his thinking and language developed.

But language is far more than a list of words. It has structure, a grammar and syntax, establishing relationships among the objects and people that the words represent. In English, much of this structure is provided by word order. The phrase "the boy saved the dog" has a very different meaning than "the dog saved the boy." A change in word order can transform a statement into a question. Prepositional phrases, such as "at the house" or "before lunch," place things in space and time. Najma noted in her diary that Zohra started using prepositions, such as "on," "over," and "under," when she was three and a half.

Children learn the words and grammar of their native language not through mechanical drills at school but from everyday dialogs and games with those around them.[7] This is true whether the language is conveyed by sounds, vision, or touch.[8] In letters to her teacher, Sophie Hopkins, Anne Sullivan describes how she threw out her lesson plans and taught Helen Keller informally, spelling into Helen's hand descriptions of all the objects and events that came their way.[9] "I am going to treat Helen exactly like a two-year-old child . . . I shall talk into her hand as we talk into the baby's ears." Najma, too, used her scrapbooks and real objects to talk to Zohra about everything that happened throughout the day. These discussions were nonstop and exhaustive. When it was time to put on their shoes, Najma might show Zohra a picture of and the word for shoes from the scrapbook and tell Zohra to get her shoes. This would lead to an extensive conversation, repeatedly incorporating the word "shoe": "There are two shoes. What color are your shoes? Let us put on your shoes," and so on.

While a small child might use one word, say "milk," to indicate a whole thought, as in "Bring me some milk," Najma always talked to Zohra in whole sentences. This, I was to read in Anne Sullivan's letters, was also Sullivan's approach. For Najma, every object, every action was an opportunity for teaching. "See the glass of water," Najma might say to Zohra. "It's only half full. Let's cut an orange in half. Now what did I just do?" Najma's family called Zohra "Najma's pocketbook" because the child was always with her. But Najma didn't mind her all-consuming work. She had no children of her own and felt that Zohra was a gift from God. Regarding Zohra, she wrote in her diary, "I always felt something was missing in my life, but now, with her, I feel my life is complete."

In her moving memoir *If a Tree Falls*, Jennifer Rosner describes a similar and just as intense approach to teaching language to her two deaf daughters, one of whom wore hearing aids and the other a cochlear implant.[10] Rosner made "who" and "where" picture books of the places and encounters her children had experienced during the week and then used the scrapbooks to narrate the events all over again. She talked constantly to her children. "You couldn't turn away, stare off, do a thousand other things. It required presence and intimacy," Rosner wrote.

Using pictures to learn the names and words for real objects and actions seems straightforward. But how did Najma teach Zohra abstract concepts such as love and justice? When I asked Zohra this question, she said it was no harder for her to learn abstractions than it is for anyone else. Later that day, I asked Najma the same question and received the same reply. Only in reading Anne Sullivan's account of teaching Helen Keller did I find a satisfactory explanation.

In her letters, Sullivan writes that she was repeatedly asked by learned doctors how she taught Helen adjectives and abstract ideas.[11] "It seems strange that people should marvel at what is really so simple," Sullivan wrote to Hopkins. "Why it is as easy to teach the name of an idea, if it is clearly formulated in the child's mind, as to teach the name of an object." Once, when Helen was working on a math problem, Anne finger-spelled onto her pupil's forehead the word "think." Immediately, Helen understood its meaning and used it correctly from that time onward. Sullivan described how children learn from experience to differentiate feelings such as happy, sad, or regretful, and then adults give them the words for these feelings. Abstract ideas such as love or justice can be learned by association: love with caresses and justice with stories. Through constant interactions and dialogs, children learn not only abstract concepts but how to behave, what is considered right and wrong, and the conventions and beliefs of their culture.

To Najma's relief and delight, Zohra said her first words, "ma," "ama" (mama), and "up," at age one and a half. But Zohra's ability to understand spoken language and to read far outpaced her ability to speak clearly. She may have had language, but she struggled to communicate with others. Her hearing, even with hearing aids, didn't provide good feedback for the sounds of her own voice. While most of us learn to speak instinctively, for Zohra it was a hard-fought effort. Najma used a mirror to show Zohra how to move her mouth to say particular words and had Zohra touch her own lips for the *m* sound and her neck for the *k* sound. Ever resourceful, Najma made yet another scrapbook, this one devoted to speech sounds. One of the hardest sounds for Zohra to make was the *k* sound, especially at the beginning of words. From her scrapbook, Najma would show Zohra a picture of a book, and they would practice pronouncing the word with its *k* sound at the end. Then they practiced "cup" and "coffee." "The day I learnt to pronounce *v* correctly for the first time," Zohra told me, "it was a personal milestone achievement. Excited about my new ability to pronounce the letter, I kept on saying it the rest of the day, non-stop, till everyone got tired of hearing it."

Zohra mastered letter sounds by age three and a half, when she was also putting three or four words together in simple sentences and questions, such as "I want milk" and "What does Papa say?" But she was almost five before people, other than Najma and other family members, could understand her speech, and then it was only friends or her teachers who knew her well. Her voice didn't sound normal. "It sort of comes from inside, from the throat," Najma wrote.

Whereas hearing children first learn language and then how to read, Zohra, thanks to the scrapbooks and Najma's teachings, learned language in large part through reading. She had already mastered a simple reading primer by the time she started school at age three and a half. This surprised her kindergarten teacher, not only because Zohra could read but also because she did not know the alphabet. Najma had taught Zohra all the letters but not their alphabetical order. Perhaps this is not so surprising since most of us learn the letter order in the alphabet through the ABC song. What's more, Zohra learned to read by recognizing words as a whole, rather than sounding out each letter. In her autobiography, Helen Keller described her way of communicating by finger spelling and how, here too, she grasped the whole from its parts: "I do not feel each letter," she wrote, "any more than you see each letter separately when you read."[12] One hundred years later, vision scientists Merav Ahissar and Shaul Hochstein echoed her thoughts in their reverse hierarchy theory: the idea that we immediately grasp the gist of a scene without first seeing the details.[13]

Zohra went to an ordinary school, not a special school for the deaf. Though deafness isolated her from other students, she liked school. She was good at arithmetic, could copy her letters, and learned the different sounds that letter combinations can make (*s* in "sweet" and "scarf"). She understood much of what was said in the classroom with a combination of listening and lip-reading and enjoyed playing outdoors with other children, where actions were more important than words. But, above all, she loved to read, a passion she shared with her grandmother. One of her favorite childhood places was "At the Green Book Shop." She was, Najma recorded in her diary, a happy and loving child who asked lots of questions.

But when Zohra was just shy of four years, she contracted an ear infection that temporarily robbed her of what hearing she had. It took six weeks to clear the infection, and Zohra missed a whole term at school. Najma worried about further infections and felt that Zohra made little progress with her language and speech at school, so she sent Zohra to school only two to three times per week and resumed her own auditory training with Zohra on nonschool days. Zohra usually didn't respond when people, out of eyesight, called her name. So, at odd moments, Najma would sneak up behind Zohra and speak her name softly, using voices of different pitches. She also went about the house with a tape recorder, recording common environmental sounds and short messages from Zohra's family and then testing Zohra on them. In the afternoons, the two would sit together on their balcony, where Najma showed Zohra a set of words. Covering her mouth, Najma spoke one of the words out loud, and Zohra had to select the correct word from the list. When Zohra mastered a short list, the list simply got longer. Najma never wasted an opportunity to teach Zohra to use what hearing she had.

When Zohra was five years old, Najma and Zohra moved to Dar es Salaam and lived in a house very close to Zohra's parents and siblings. Her two brothers and sister called Zohra "if, why" because of all the questions she asked. By the third day of school, Zohra, always interested in other people, knew the names of every student in her class. She had made a list. Although she sometimes felt isolated from other students, who could talk and socialize so effortlessly, she developed several close friends, forming a small social circle, which was, to her, like family. Najma worked in Zohra's school, first as a librarian and then as a classroom teacher, though not in Zohra's class. Since Najma had taught at a religious center since age eighteen, she was comfortable with teaching, but she gave it up after a short while. Teaching required too much talking during the day, and Najma spent her evenings talking nonstop to Zohra.

Despite Najma's efforts, by age eight Zohra's speech had deteriorated, becoming more nasal. This change, Najma suspected, occurred because Zohra's hearing had gotten even worse. So Najma took Zohra, now age nine, to the United States, to the Clarke School for Hearing and Speech in Northampton, Massachusetts. Zohra received better hearing aids, but they didn't help much, and her speech was already better than that of most of the kids at the school. They only stayed one semester.

After returning from Northampton, they moved back to Moshi, back with Najma's parents (Zohra's grandparents), where the family had a mattress store. Zohra went to the international school, where the classes were taught in English. She loved learning new things, so when she got home, she would sit her grandmother down by the blackboard in their house and teach her. Zohra was lucky that her grandmother knew English, since most of the older women in the community did not. Both Zohra and her grandmother were avid readers, so they talked about their favorite books, and her grandmother shared her own life stories. Since Zohra's grandmother did the cooking, Najma had time to work with her father in the store. She was able to build up the business, which she loved doing. But her free time was still devoted to Zohra. "Close the door and work with Zohra," Najma's dad would say, and Najma did—for three hours per day.

Still, Zohra heard less and less, and as a result her speech continued to worsen. At about this time, Najma's uncle read an article about the cochlear implant, a device different from hearing aids, which could help the profoundly deaf. Once again, Najma took Zohra, now twelve, to London, back to Portland Hospital, where her suspicions were confirmed. Zohra's hearing loss had increased; she couldn't hear sounds below 110 decibels. Hearing aids were of little use. Only a cochlear implant could help her hear—but it cost close to $50,000.

By the time they returned to Moshi, Najma was despondent. She told her uncle that there was no hope. The only possible solution for

Zohra's hearing was well beyond their grasp. But Najma's uncle took matters into his own hands. He sent an email to the extended family, everyone contributed, and Najma's brother secured a loan for $10,000. Funds in hand, Najma and Zohra traveled overseas once more, this time to Toronto, Canada, where Zohra's cousins lived. The implant was a little less expensive there. The cost was $48,000.

– chapter 11 –

Persistence Pays Off

In 1790, Alessandro Volta (for whom the volt was named) stuck one end of an electrical circuit into each ear and experienced "a boom within the head," followed by a sound similar to that of "boiling thick soup."[1] Given this shocking experience, he never repeated the experiment, but his observations provided the first evidence that our perception of sound is mediated, in part, by electrical events in our head.

Nearly two centuries later, in 1957 in Paris, André Djourno and Charles Eyriès performed the first direct stimulation of the human auditory nerve. Eyriès, an otolaryngologist, was operating on a man who had lost both cochleas, the hearing organs in the inner ears. During the operation, he placed a coil of wire, designed by Djourno, on the cut stump of the auditory nerve, the nerve that leaves the inner ear for the brain. An external microphone was used to pick up and convert ambient sounds into electricity, which was then transmitted to the coil on the auditory nerve via an external induction coil on the outside of the patient's head. With this device, the patient could hear environmental or ambient sounds.[2]

Sound is created when objects vibrate. As they oscillate, they push (compress) and pull (rarefy) the air around them, causing pressure waves that radiate outward. If these pressure waves are low in frequency, less than twenty cycles per second, we feel them as vibrations on our skin. But if their frequency lies between twenty and twenty thousand cycles per second, we hear them as sounds. Sounds in nature are almost always complex: each sound is made up of a mixture of many pressure or sound waves of different frequencies (which we hear as different pitches) and amplitudes (which we hear as different degrees of loudness). These pressure waves vibrate our eardrums, which set in motion the three bones of the middle ear, which push, in turn, on the oval window of the cochlea, the structure in the inner ear that is shaped like a snail shell. This sets in motion fluid in the cochlea, which pushes on the basilar membrane, stimulating the inner hair cells, our sound receptor cells. Inner hair cells convey their message to neurons in the auditory nerve in the cochlea, which pass the information along the auditory pathway to the brain. All these steps help to amplify the effects of even low-amplitude sound waves. Djourno and Eyriès's patient had no functioning ears, so they provided their patient with sound by stimulating his auditory nerve directly.

By sheer luck, a patient in Los Angeles showed his ear surgeon, Dr. William House, a newspaper article about Djourno and Eyriès's work.[3] Dr. House was intrigued. Many people lose their hearing when the majority, if not all, of their inner hair cells die. Could he restore hearing to the deaf by developing a device that bypassed the ear entirely and stimulated the neurons in the auditory nerve instead? In 1961, he provided two patients with the very first cochlear implants. These implants worked in a similar manner to the device made by Djourno and Eyriès except that the stimulating electrode was placed not on the auditory nerve itself but in the cochlea. Ambient sounds led to the firing of the cochlea electrode, which excited auditory neurons in the center of the cochlea. They sent their message along the auditory nerve to auditory areas in the brain, enabling the patients to hear.

However, redness and swelling developed around the internal coils, and Dr. House, fearing that the patients' bodies were rejecting the devices, removed them. Given his heavy surgical responsibilities and young family, Dr. House couldn't work further on cochlear implants for several years. But in 1972, after developing a portable implant, he sent one patient, Charles Graser, home with the device. The electronics for this device were worked out by his longtime collaborator, Jack Urban. An external stimulator provided an alternating current whose amplitude was modulated up and down by external sounds. This current was then transmitted via radio waves to the electrode in the cochlea, which stimulated auditory neurons. On the first full day with his cochlear implant, Graser, who lost his hearing as an adult, enjoyed the many sounds he heard on his morning bike ride and was able to understand some speech in quiet surroundings.

Meanwhile, additional groups of scientists, engineers, and physicians began developing cochlear implants, in both the United States and Europe.[4] And half a world away, in Melbourne, Australia, Graeme Clark embarked on his own design. Both House and Clark wrote memoirs detailing their work on the cochlear implant.[5] Although the two surgeons disagreed on the number and length of cochlear electrodes, their memoirs share much in common, a sentiment summed up by the title of House's book, *The Struggles of a Medical Innovator*. While developing their implants, both surgeons met with a huge amount of hostility from their medical colleagues, and both struggled to find funding to keep their research programs alive. Just as Sir Harold Ridley faced opposition with the intraocular lens, many of House's and Clark's colleagues balked at the idea of introducing a foreign object into a delicate sensory organ. And objections to the cochlear implant went much further. In 1964, Dr. Merle Lawrence, an authority on hearing, stated, "Direct stimulation of the auditory nerve fibers with resultant perception of speech is not feasible."[6] Some investigators were concerned that auditory neurons in the deaf may have already degenerated from disuse so that the implant would have few neurons to stimulate. Others predicted that the stimuli provided by the implant would result in perception so crude as to be useless for recognizing environmental noises, much less speech. They accused House of misleading patients who were desperate to hear and implied that he was doing his "human experimentation" just for the money. (Yet House never took out a patent for his designs so that his developments would be freely available to all.) By 1975, thirteen people in the United States, most of them patients of Dr. House, had functioning cochlear implants. The National Institutes of Health commissioned a study of these thirteen individuals, and the investigation, published in 1977, reported that the patients benefitted from the devices. At that junction, medical and scientific opinion began to change.[7]

But resistance to the cochlear implants came from another source, the Deaf community.[8] Imagine what it would be like to lose your hearing as an adult. Although you've communicated most of your life with the spoken word, you can no longer hear what others say. Unable to interact easily, you feel cut off from society. If, however, you were born deaf, you may have grown up in the Deaf community, communicating with others through one of the sign languages (Sign). (As with spoken languages, there are different sign languages in different regions and countries.) As a deaf child, you grew up in a community and culture in which you belonged as an equal member but also with the knowledge that your community was continually threatened. Over the past two centuries, the deaf have been forced to fight hard for their language and way of life.[9] Particularly in the nineteenth and most of the twentieth centuries, many educators in the hearing community insisted that deaf children learn an oral language, even going so far as to forbid sign languages in their schools. Since it's extremely difficult to learn an oral language when you can't hear, many deaf children were deprived of useful communication. Here was a case involving educators, who could hear, thinking they understood deafness better than the deaf themselves. Could the introduction of the cochlear implant be another example of forcing oral language upon the deaf? Could a person with a cochlear implant, which provides subnormal hearing, feel the equal of a person with normal ears? What would happen to the Deaf culture and Sign, a community and language that sustain the deaf, if most deaf children were implanted with the device and educated to live in the hearing world? These objections and concerns from the deaf reached their peak in the 1990s, when cochlear implants became more common, and are still with us today.[10]

Both House's and Clark's memoirs detail the dogged persistence that the two men possessed in overcoming not only political but also technical problems presented by the cochlear implant. These challenges were never far from their minds. The cochlea is arranged "tonotopically." Hair cells and auditory neurons at the base of the cochlea (near the junction with the middle ear, the fat end of the spiral) respond to high frequencies, while those at the apex respond to low frequencies, with a gradient of frequencies in between. To maximize the frequencies his patients could hear and distinguish, Graeme Clark hoped to introduce an electrode array that could extend through much of the spiral-shaped cochlea.[11] In this way, different electrodes on the array would stimulate different regions of the cochlea. If the electrode array was too flexible, however, it would collapse as it encountered a turn in the cochlea, while a stiff array might not bend around the spiral. This problem nagged at Clark until one day, while on vacation at the beach, he had an epiphany. He picked up some spiral snail shells and tried introducing different types of grass stems and twigs into them. He discovered that a stem that was flexible at the tip but stiffer at the base worked well. The flexible tip allowed the stem to bend around the turns, while the stiffer base prevented the stem from collapsing. To this day, the electrode array introduced into the cochlea has a gradient of stiffness, lower at the tip and greater at the base, and extends about one and a quarter turns into the two-and-a-half-turn spiral cochlea.

The various models of the cochlear implant all work along the same principles. A microphone, located in the headpiece behind the ear, picks up ambient sounds and converts them to electricity. These signals are sent to the speech processor, which digitizes and separates the sound into different frequency bands. After further processing, the signals are sent back to the headpiece and transmitted via radio waves to an internal computer chip inside the skull, which, depending upon the signals it receives, selectively fires different electrodes along the cochlea. Electrodes located near the base of the cochlea excite auditory neurons selective for high-frequency sounds, while electrodes located further toward the apex stimulate low-frequency neurons. The speech processor usually has the same shape as a hearing aid and, likewise, sits atop the ear, but some manufacturers build it into the headpiece that holds the microphone and transmitter coil. Waterproof versions of the speech processor can be worn on the arm. Magnets, both on the headpiece and on the part of the implant located inside the skull, keep the internal and external parts in apposition to each other.[12]

One great advantage of this general design is that the cochlear implant can be upgraded by modifying the computer programs in the speech processor, thus eliminating the need to replace the internal parts of the implant. Each patient with a cochlear implant undergoes many "mapping sessions" in which the settings in the speech-processor program are adjusted for their most comfortable and useful hearing.

In a healthy cochlea, three thousand inner hair cells, arranged in a tonotopic manner, stimulate auditory neurons. A cochlear implant contains, at present, a maximum of twenty-two electrodes, and not all these electrodes can be used at one time to stimulate circumscribed regions of the cochlea. Since the ratio of inner hair cells to electrodes is more than one hundred to one, it's not surprising that people with cochlear implants cannot discriminate sound frequencies as well as people with normal hearing. They may not hear, for example, the difference between the notes C and C-sharp played on the piano. What is surprising, however, is how well a person with a cochlear implant can hear. Many understand speech without lip-reading, enabling them to talk on the telephone, and those who receive implants in both ears are able to understand speech in noisy environments and to localize sound. Many children born deaf but implanted in the first year of life manage well in a hearing world. The remarkable and surprising fact that a cochlear implant works so well is a testament to the adaptability and plasticity of the human brain.

– chapter 12 –

An Uncanny Feeling

BY THE TIME ZOHRA WAS ABOUT EIGHT YEARS OLD, HER HEARing aids no longer helped. She heard nothing at all—not voices, footsteps, car motors, or even the low, rumbling white noise she used to hear through her hearing aids. By age twelve, she had forgotten what hearing was like. So when her cochlear implant was first turned on, in the spring of 2000, and a beep was sent through the device, she didn't realize she heard a sound. Certainly, she sensed something—a feeling in her head, something uncanny and uncomfortable. Zohra knew she was supposed to hear beeps, so when a second beep was sent through her implant, and she experienced that weird sensation again, she thought she must be hearing. That explained, too, the strange feeling she experienced when she looked over at Najma and the audiologist and saw their lips moving.

Zohra was hearing, but it was not the hearing that most of us experience. It was a paradox. She wrote, "I remember walking along the street in downtown Toronto, after leaving the hospital, I was 'hearing' . . . the cars and other sounds, but I could not recognize them or make sense of them, but I was 'hearing' them." In a similar vein, Oliver

Sacks titled his story about Virgil "To See and Not See" because Virgil gained sight but did not understand what he saw.[1]

The next day, when Zohra put on the implant, the sounds were "loud, scary, and uncomfortable." It was hard to reconcile these sensations with what others found so meaningful. While most of us tune out background noise, often without being aware we are doing so, Zohra was bombarded by sounds. She did not want to wear the implant. But Najma, her cousins, and her grandparents all insisted, and after two weeks, the sounds she heard were no longer so unsettling. There are still moments, even today, when Zohra removes her implant to get rid of the noisy distractions. But gradually she began to feel more connected to the world and other people.

Zohra's initial experience with the cochlear implant is one most of us can hardly imagine. But we can recall how we felt when we sensed something unexpected. We may have walked casually down a flight of stairs under the assumption that every step was the same size. As we stepped down, however, the lower step was a further drop than we anticipated. We felt ourselves untethered in space, momentarily disoriented, and this triggered a whole series of heart-pounding, stomach-tightening, visceral reactions that lingered even when we reached the lower step safely. I felt these types of sensations many times when first learning how to use my eyes together and see in 3D—moments of disorientation and alarm. Yet I was experiencing a qualitatively new way of seeing, not sight for the first time. Hearing for the first time in memory, as Zohra experienced, is much more disruptive and unsettling. She heard sounds that had no meaning; they did not fit into her concept of the world.

At what age does an infant with normal senses separate sounds from sights and from the feeling of being touched? After all, we perceive events, not individual, isolated stimuli, and almost all events are multisensory.[2] A baby tastes, smells, and feels the viscidness and warmth of her mother's milk; she sees her mother's face while hearing her voice. Since sensory areas in the cortex may not be as specialized in babies as they are in adults, infants may experience the whole of an event without distinguishing sights from sounds.[3] In *Descartes' Error*, neurologist Antonio Damasio writes, "In the beginning, there was no touching, or seeing, or hearing, or moving alone by itself. There was, rather, a *feeling of the body* as it touched, or saw, or heard, or moved."[4] As we mature, we not only see the external world but also become conscious of ourselves seeing.[5] Through active exploration of our environment, typically in infancy, we become aware of ourselves sensing the world and the different organs and modalities that we use to sense it. When her cochlear implant was first turned on, Zohra not only heard but became aware of herself hearing.

– chapter 13 –

Squeaks, Bangs, and Laughter

WHEN WE HEAR SOUNDS, WE ARE HEARING EVENTS. WE HEAR what causes the sounds. If nothing is happening, if the world is absolutely still, there is nothing to hear. This was perhaps best understood by John Hull, a blind man who, in *Touching the Rock*, describes what it was like not to see and thus to sense the world primarily through hearing: "The strange thing about it, however, is that it was a world of nothing but action. Every sound was a point of activity. Where nothing was happening, there was silence. That little part of the world then died, disappeared."[1] Before her cochlear implant, Zohra was only aware of events she could see. To be able to hear things she couldn't see was an uncanny experience, both fascinating and unnerving, as would be the case if we were suddenly given the ability to see through walls.

Zohra found it easiest to learn sounds that were produced by her own actions. She pushed a chair and heard it squeak across the floor. In her memoir, *If a Tree Falls*, Jennifer Rosner describes the moment her daughter's cochlear implant was first turned on.[2] The little girl was given a drum and drumstick. She whacked the drum (who wouldn't?) and was startled. She whacked it again and laughed. What a brilliant way to turn on the cochlear implant! Give the child something to act upon so that she knows what causes the resulting sound and where it's coming from. Intuitively, we have known this all along: giving our babies a toy rattle to play with serves the same purpose.

While riding in a car, Zohra couldn't initially distinguish a person's voice as a separate sound from the rumble of the motor. They seemed to be all one sound. Indeed the sound waves produced by the car and the person arrived at Zohra's cochlear implant (or at normal ears for that matter) all at the same time. Zohra put this so well when she wrote to me, "Today when I was in the elevator with my mum, it was amazing how I could smell two things at once, perfume and cigarette smoke. I could separate the two smells easily. Focus on one at a time. They were distinct. And I told Mama, I can't do that with hearing. All the sounds are mixed together, and it's so hard to separate them into their own parts like layers."

Just like Liam had to learn to what objects different contours and colors belonged, Zohra had to learn to what sources different sounds belonged. This is made more difficult because each source—a motor, a person's voice—is made up of numerous sound or pressure waves of different frequencies and volumes. How do we group the sound waves that belong to a voice and segregate them from sound waves that belong to the motor? While Liam could study a single view for some time, sounds are ephemeral. So Zohra could teach herself sounds most easily if they resulted from her own actions, which could be repeated again and again.

Like object recognition, sound recognition depends upon communication between lower and higher sensory areas in the brain. We are only beginning to discover where along the auditory pathway sound processing takes place. Neurons in the auditory nerve connect with neurons in the cochlear nucleus in the brainstem, which connect in turn with neurons in other brainstem nuclei, the midbrain (inferior colliculus), and the thalamus (medial geniculate bundle) before reaching the primary auditory cortex located in the temporal lobes.[3] While the primary *visual* cortex (V1) is organized in a *retinotopic* manner (adjacent areas of space are mapped to adjacent areas of the primary visual cortex), the primary *auditory* cortex is organized, like the cochlea, in a *tonotopic* manner (tones close to each other in frequency are processed in neighboring regions of the primary auditory cortex). This neural attention to frequency helps with sound source recognition. Drop a spoon on a tiled kitchen floor, and you know immediately whether the spoon is metal or wood by the high- or low-frequency sound waves it produces upon impact. The primary auditory cortex is encircled by and communicates with higher auditory cortices. As with vision, neurons lower in the auditory pathway respond to the raw features of the input, which for audition include changing frequencies over time. Neurons higher in the auditory pathway respond more selectively to categories of sounds, such as bird calls or voices.[4] Just as we can speak of visual object recognition and its loss with visual agnosia, we can speak of auditory object recognition and its loss with auditory agnosia. People who have intact primary auditory cortices but have lost higher auditory areas still hear sounds but, like Zohra when she first heard through her implant, have trouble distinguishing the source of one sound or isolating one auditory object from another.[5]

Albert Bregman called our ability to organize sounds by their sources "auditory scene analysis."[6] His experiments and those of others reveal grouping rules for sound waves that resemble Gestalt rules for visual perception. For example, we group sound waves of similar frequencies together. We hear these as similar pitches and perceive them as coming from the same source. If sound waves of different frequencies start and stop together or grow louder and softer together, they probably come from the same source. In an analogous way, we see and recognize the different parts of an object as a single unit because they all move together.[7] The reverse hierarchy theory, described earlier for vision, applies to audition as well.[8] According to the theory, we see and recognize an object as a whole unit before we are aware of all its features, and we instantly recognize the source of a sound without being aware of all its individual sound waves. As with vision, we may only be aware of sounds after the information has swept up the auditory pathway and been processed in higher auditory areas. With so little hearing in childhood, Zohra's auditory system never fully developed. As she began to differentiate and recognize sounds, new connections likely formed within and between the auditory areas in her brain.

Zohra was not shy about asking others what she was hearing, and she carefully and tirelessly analyzed each new sound. She figured out that the unnerving swish she heard when moving was the sound of her own clothes rubbing against her. She expected footsteps to make noise but was delighted to learn that she could tell the type of shoes people were wearing by the sound of their footsteps. Sounds, she thus discovered, reveal the material properties of objects. The way some things sounded surprised her. While eating potato chips after receiving her implant, Zohra not only experienced the snap of the potato chip but also the sound of her own chewing. She wrote, "I found it so weird that the softest things made so much sound. They were so loud and I had never imagined them to be like that. To me, potato chips were always such a delicate thing, the way they were so lightweight, and so fragile that you could break them easily, and I expected them to be soft-sounding. But the amount of noise they make when you crunch them was something out of place. So loud." The wind felt soft on her face but made a lot of swishing noise. Paper may be delicate but made a loud crackle when crunched. The keys on a computer keyboard felt light to the touch, but typing made a racket. Brushing teeth, sweeping the floor, putting a key in a lock—all of these made different sounds. When she dropped something, she was surprised and happy to learn, the object made a sound when it hit the floor.[9] The same thing doesn't always sound the same. Water, when coming out of a tap or boiling on a stove, makes different sounds. (In a similar vein, Liam was surprised that hot water coming out of a tap looked opaque, not transparent as he expected.) Since Zohra must take her implant off when showering, I asked her if she recognized the sound of water going down a drain. She wasn't sure. At that time, in 2010, we were talking in my office where I had a large sink. So we filled it with water and then listened to the water drain out. Zohra was delighted. "Want to do it again?" I asked. Zohra smiled and nodded, so we filled and drained the sink for a second time.

When Najma and Zohra returned to Moshi after six months in Canada, Zohra discovered and relished all sorts of new sounds, everyday sounds that most take for granted. Zohra recalled in an email,

> One thing I remember the most is that at home in Tanzania we have something called a mosquito mesh which is a metal screen that we place on the doors and windows to keep the mosquitos away. They are like a metal net in the form of a sheet, very flexible and soft. When I got home I would always hear this sound, especially during nighttime where other environmental sounds were quiet. It was like a banging sound but softer and dampened. And I would ask my family where the sound was coming from, and they would say it is the windows. Whenever they said that, I wasn't happy with the answer because the windows weren't moving. They were shut so how could they be banging? One day when I heard the sound again, I opened the curtains and went to have a look at the windows myself and saw that whenever the metal sheets moved with the wind, the sound came.
>
> And there's this distinct sound that remains so vivid in my memory. It's the sound of the thick broom brushing against the floor. At home early mornings we would always clean our floors with soap and water and for that we would use a small broom with very thick wooden bristles—similar to a heavy duty garden/garage broom here, but smaller, and I would hear that sound every morning.

And Zohra discovered one of nature's most beautiful sounds: "Watching the rain was always my favorite thing to do. I remember watching the rain from the backyard at home. I could see the rain drops and the water hitting the roof and pouring down. And I loved the smell of the rain, and I felt the cool mist on my face. And now with the implant, I could hear the rain; I could hear it hitting the roof. I always used to see, smell and feel rain. Now I heard rain."

All through Zohra's childhood, Najma had tried to teach her to pay attention to sounds. But it was so easy for Zohra to ignore what little she heard and depend upon her sight. With the cochlear implant, Zohra learned that most actions cause sounds. She began to expect this, and expectation plays an important role in perception. When she opened a cardboard box, for example, she knew the movement of the lid would make some noise. "A sound," Zohra wrote, "is not a noise anymore but more like a story or an event." As Zohra learned the meaning of sounds, they became less threatening. Indeed, they made her feel safe.

One day, Najma casually asked Zohra if she had been playing the word game Scrabble earlier that morning. Indeed, Zohra had opened the box, spilled out the letter tiles, and played by herself. From the sound of the tiles tumbling out of the box, Najma, listening from an adjacent room, had inferred that Zohra was playing the game. That a simple sound could imply so much made a big impression on Zohra. Now, she really began to listen.

"WHAT IS YOUR FAVORITE SOUND?" I ASKED ZOHRA WHEN WE first began talking in 2010. Without hesitation, she answered, "Laughter." The sound of laughter came to her as a complete surprise. While she knew oral speech was made up of sounds, she didn't know the same was true for laughter. Although she loved to laugh herself, she thought laughs were all visual. But now that she could hear laughter, it seemed more real, deeper, and highly infectious. She ended up laughing too.

At first, all sounds were frightening to Zohra because they were meaningless. But now, only angry sounds, like a door slam or an irate voice, are scary. Sounds that others find irritating, such as fingernails rubbing on a blackboard, she finds irritating too. Over and over again, Zohra emphasized the emotional content of sound. She found the rhythm and cadence of human speech to be soothing, just as a baby is calmed by the sound of her mother's voice. As we were speaking in my office, a number of students were talking together in the hall. I could understand what they were saying, but Zohra could not. She could, however, discern the emotional tone of their speech and found their voices comforting. From the sound of others, she knows now, without even looking, that she's not alone in a room. It's like having "another eye," Zohra told me. It gives her a sense of belonging.

Zohra's discovery of the emotional impact of sounds contrasts sharply with Liam's description of his first sights. Most sights did not arouse strong emotions, and he was confused by what others found to be beautiful or ugly. Anthony Storr, in his brilliant book *Music and the Mind*, writes that there is a closer relationship between hearing and emotional arousal than there is between seeing and emotions.[10] His thoughts are supported by laboratory tests indicating that people are better at judging the emotions of others from hearing their voices than from watching them behave.[11] The power of sounds to manipulate our emotions is certainly exploited by moviemakers, who add music to enhance the impact of their visual scenes. Storr recalled a friend who found his first view of the real Grand Canyon to be less dramatic than views of it he had seen in movies. The films were always accompanied by music that added to the drama. A view of a wounded animal or suffering person may produce less of an emotional response than hearing the wounded creature scream. The sound of the human voice may be comforting but may also be harnessed in a deadly, manipulative manner. Hitler used the hypnotic qualities of his voice to be persuasive. On a more positive note, laughter therapy works in part because the sound of laughter, as Zohra discovered, brings forth positive emotions.

It's not surprising that sounds evoke strong emotions because they provide an essential and important early-warning system. We can hear sounds in the dark, around corners, and from up to six-tenths of a mile or a kilometer away. We produce sounds that evoke strong emotions, such as growls, sobs, and laughter, and we respond more quickly to these nonspeech sounds than we do to words.[12] We squirm at the sound of fingernails rubbing on the blackboard because this action produces a pattern of sound waves similar to those produced when people scream.[13] Laughter may trigger happy feelings in part because it causes the release of naturally occurring opioids in the brain.[14] The auditory cortices, as well as subcortical auditory areas, are interconnected with structures, such as the amygdala and nucleus accumbens, that process the emotional meaning of events.[15] As Zohra discovered, we are wired to respond rapidly and emotionally to sounds.

A lack of sound also has a strong emotional impact. Zohra, after a childhood of deafness, is comfortable in a silent world and, indeed, will remove her cochlear implant when sounds become overwhelming. Much more frightening to her is total darkness. But most of us are unnerved by silence. Although we may be hardly conscious of background sounds, we are also dependent upon them for our emotional well-being. A silent world is a world without action, a dead world, and most of us find it frightening.

– chapter 14 –

Talking to Others

WHILE IT WAS A GREAT ACCOMPLISHMENT TO RECOGNIZE environmental sounds, what Zohra really wanted was to talk freely with others, to understand speech and be understood. To this day, words are much harder for Zohra to identify than most environmental sounds. She describes environmental sounds as "more solid." She can imagine the purr of a car motor but not the sound of a whole sentence.

Because we communicate, in large part, through speech, we must extract more information from spoken language than from most environmental sounds. We may hear a car go by, but we don't usually have to know what kind of car it is. With speech, however, we must recognize the subtle differences in the way different words sound. Small but important changes in the way they are spoken tell us if we are hearing a statement or a question or if the speaker is happy, angry, or bored.

After her operation, Zohra's surgeon told Najma that he would not have given Zohra a cochlear implant had he known just how long she had been profoundly deaf. Zohra was prelingually deaf, meaning she was deaf before she developed speech. With her hearing aids, Zohra may have picked up the sounds of some of the consonants that are hard to lip-read, such as *h*, *g*, and *k*, which would have helped with understanding speech, but throughout her childhood, she may not have heard distinctly a single word. So Zohra faced an even tougher challenge than people who are postlingually deafened and receive a cochlear implant. They may remember what words sound like. Though speech through a cochlear implant initially sounds electronic and robotic, many postlingually deafened adults can begin to understand spoken sentences in quiet surroundings shortly after receiving their implant.

After her surgery, Zohra remained in Canada for six months in order to receive speech therapy. She had therapy in the morning, and then Najma would drill her throughout the rest of the day. Zohra's first task was to differentiate speech from nonspeech. Fortunately, spoken language has some distinguishing characteristics. In English, for example, syllables are spoken at a rate of two to ten per second, and the pitch and loudness of a given person's speech usually varies over a relatively narrow range.[1]

We naturally teach infants oral language with "motherese," a slowed-down, simplified version of speech with exaggerated emotional content. But Zohra, from the start, had to learn the faster, more complicated speech of adults. This was difficult and exhausting, and sometimes both Najma and Zohra ended up in tears. After one or two months, Zohra could tell the speech of men from that of women. The very first word she recognized was "banana." Vowels, she discovered, were easier to hear than consonants. Phone numbers and other number sequences, recited on the radio, were easy to pick out, perhaps, Zohra thought, because of the rhythm of the utterance. Arlene Romoff, who wrote about her experiences with a cochlear implant in *Hear Again*, also found that number sequences popped out.[2]

Zohra worked on distinguishing different phonemes, or the smallest unit of speech that can make one word sound different from another. The words "cap" and "cup" differ by just one phoneme, and Zohra had difficulty telling them apart. Because of her deafness, she missed the window early in life when infants tune up their sensitivity to phonemes. Each language makes use of a subset of all the possible phonemes we can hear. Newborns may hear all these phonemes equally well, but within a year of listening to their mother tongue, they become most sensitive to the phonemes of their native language. Thus, most native English speakers have no trouble hearing the difference between the sounds made by *r* and *l*, while this difference is hard for a native Japanese speaker to hear. On the other hand, English speakers have a hard time distinguishing the difference between the sounds produced by *u* and *ü* in German.[3]

At first, Zohra worked with "closed sets" of one-syllable words. After covering her own mouth, Najma or the therapist would say a word, and Zohra would have to pick it out from a list of two, then three, then up to fifteen words. Once she mastered this, Zohra moved on to "open-set" activities where there was no list of words to choose from. The possibilities were endless. Zohra was quizzed on common phrases such as "What is your name?" or "What is the weather today?" Eventually, she listened to whole paragraphs from books and was quizzed on her comprehension. She listened to words spoken from increasing distances or in quieter voices. She even learned to understand whispered speech and to whisper herself, a skill she could never have developed without the cochlear implant.

More than a decade later, when Zohra was in medical school, her family adopted an African gray parrot, a beautiful talking bird. Now Zohra could hear the way another being learned to talk and how her family members spoke "motherese" to the bird, often repeating phrases over and over again. And Zohra never stopped enjoying the way the parrot talked (usually nonstop, throughout the day), especially when the bird called her name—"Zohraaaaa!"

Context plays an enormous role in speech comprehension. When we hear the beginning of a given phrase, we can often predict how it will end. And we know what words might be used if the discussion is about a particular subject. Here, like in vision, "top-down" influences may facilitate our understanding of speech. Many years after receiving her implant, Zohra, fresh from her medical school education and now living in Canada, was listening to the radio and thought she heard the broadcaster mention the acronym "DVT," a medical term standing for deep vein thrombosis. But when she heard the rest of the broadcaster's sentence, ". . . and Highway 401," she realized that the acronym was not "DVT" but "DVP," which stands for the Don Valley Parkway, the name of a highway near her Toronto home. It is still hard for her to hear the difference between the *t* and *p* sounds, but context, in this case, provided the necessary clue. Being such an avid reader, Zohra has great fluidity with English, and this helps her to predict and interpret a sequence of words. But with reading and writing, there are obvious gaps between each word. These gaps are much less evident when we speak, a fact apparent to me when I learned French in high school. The native speakers all seemed to speak too fast.

In her memoir *Hear Again*, Arlene Romoff beautifully describes different stages in understanding speech, stages she labels with reference to the Olympic medals: bronze, silver, and gold.[4] With "bronze" hearing, you recognize words as words but cannot understand their meaning. The words sound like English but make little sense, so to follow speech, you remain very reliant on visual cues such as lip-reading. With "silver" hearing, you hear the words and can access their meaning without visual cues. But you have to hear the words first and then figure out their meanings, so speech recognition is slowed. At the "gold" level, you hear and understand the words in a single instant. No extra effort is involved. You simply hear and understand. For a cochlear implant user, the quality of the acoustics can make a critical difference in the hearing level. I think I reached the silver level with my high school French, and Zohra's speech comprehension lies somewhere between the silver and gold levels.

Language comprehension, like object recognition, requires far more than the reception of raw sensory input; it demands more processing than provided by the primary auditory cortex alone. Indeed large areas of the brain, beyond the auditory cortices and involving areas of the frontal and parietal cortices, are involved with the understanding and production of speech. People who have intact auditory cortices but have lost higher language areas still hear sounds but suffer from different forms of aphasia, an inability to understand language or produce speech. Zohra's cochlear implant provided her with an abundance of new sounds, but, like Liam with visual object recognition, she had to develop higher sensory pathways to begin to comprehend speech.

A cochlear implant does not provide the range of or sensitivity to pitch provided by normal ears. Fortunately for Zohra, pitch does not contribute greatly to speech comprehension, at least where English and other Indo-European languages are concerned. We can understand the same utterance delivered by the high-pitched voice of a child or the low-pitched voice of a grown man. It would have been more difficult for Zohra to have learned Mandarin or other tonal languages in which pitch contour plays a greater role in the meaning of words.

Zohra came up with her own teaching strategies, making great use of closed captions on videos and TV. She wrote,

> All my life, before the implant, I had been reading. It was so exciting to see how those words sounded like and to perceive them in another form besides reading (and lip-reading). I would watch almost anything on TV, even the boring documentaries, just so that I could read and hear the words simultaneously. It brought me joy. Sometimes, the closed captions were lagging behind, so I would test myself often. I'll hear a phrase or bits of it, and wait for the captions to show up, and see if I got the words or phrase right. And whenever I did, I was so happy. And today, I still continue to do that.
>
> Till today, I still love reading. I love the smell of books. I loved school libraries and used to borrow so many books. An Indigo (like Chapters) store opened recently near my home. After stepping inside, I told my friend, I am in heaven.

Zohra not only enjoyed reading English; she enjoyed reading French. Back in Tanzania, one year after receiving her implant, she had to take French classes. She hated the oral lessons but loved to read and write in the language. Indeed, some of her best grades in middle and high school were in French.

While it is hard for Zohra to recognize individual voices, she can always recognize Najma's, the voice of the person she is closest to. And knowing her voice led to some surprising discoveries. Najma's voice sounded hoarser when she first woke up than it did later in the day. Once on the phone, Zohra wondered why Najma's voice sounded funny until she realized that Najma had a cold. Voices sound very different in the bathroom or other places with lots of reverberations. Zohra would speak out loud in the bathroom to practice hearing a voice in the presence of echoes.

With hearing, Zohra told me, "You have to expect a certain amount of variability." When I asked Zohra to explain this further, she used a visual analogy, accurately describing some of the challenges that Liam continually faces:

> We see the sky every day, yet it is not the same sky that we saw yesterday or that we will see tomorrow. The colors, cloud pattern, and lighting are different. Yet we recognize and see it as a sky. Not as a different object or something else. When I see an object with different lighting, angles, and shadows, the shadowing does not confuse me or make it difficult for me to perceive the object.
>
> The same goes for sounds and spoken speech . . . We recognize words and sentences such as "Hello" when they are spoken in different voices, accents, volumes, and pitches, and even in different locations . . . Have you ever had a conversation with someone on the stairwell? Hearing people find it easy or they hardly notice the effect that the echo has on speech perception. It does not make any difference to them like how shadows and lighting make no difference to those who are able to see. When I see an object with shadows and lighting, I do not even think about it, and I can recognize the object easily and effortlessly. But I find it difficult or impossible to hear sentences on the stairwell. To me the sound appears transformed or distorted, to a "new word," that I do not

understand, and the "Hello" is not the same as how I hear it in a quiet room without any echo effect.

Since the world and everything in it is constantly changing, it's surprising that we can recognize anything at all. Liam and Zohra both had to learn the parts and properties that don't change. There are invariances in the physical world.[5] The sun, for example, always shines from above, causing shadows and shading in predictable patterns that we use to interpret depth. Then there are properties unique but constant for each individual. Although a person's face may change with each expression, the overall arrangement of and relationship between the features remain the same. A person's voice may vary with mood, but it has a certain quality, rhythm, and accent all its own. We instantly recognize a friend's footsteps, gait, or handwriting, even when it is hard to dissect just how we do this.

To this day, Zohra relies on a combination of hearing and lip-reading to understand speech. Lip-reading alone is not sufficient, and she feels at a loss without her cochlear implant. Since the implant doesn't deliver the sensitivity to pitch that most of us enjoy, it is hard for her to hear the upward inflection in a person's voice when they ask a question. To her, "I go to the store," sounds the same as "I go to the store?" But when I said this phrase in the normal and then questioning way, Zohra laughed and said she could see the question on my face.

David Wright, who was deafened at age seven, writes in his memoir *Deafness* that deaf people do not necessarily see more but see differently.[6] Imagine, for example, watching and listening to someone about to finish a phone conversation. A hearing person will be aware that the call is almost over by the speaker's words and tone, while a deaf person might note a change in the speaker's stance, their head drawing ever so slightly away from the phone, a subtle shuffling of their feet, and a change in expression indicating a decision has been made. Lip-reading, Wright points out, should more accurately be called face reading or speechreading. He had a much harder time lip-reading if the speaker's eyes were hidden by sunglasses.

In 2020, when coronavirus spread around the world, everyone was instructed to wear face masks to reduce the spread of infection. Zohra was seized with anxiety. How would she cope? How could she understand other people without seeing their mouths move as they spoke? To her surprise, she found it easy to understand her family members even when they wore their masks. She had gotten used to their voices. But her experience at work was more variable. She understood people who enunciated clearly, but she struggled to follow those with strong accents. Common phrases weren't hard to pick up; nor were conversations around a familiar topic. But Zohra found it much more difficult to follow a person's speech if she couldn't predict what they'd be talking about. So she'd pull out her notepad and ask the speaker to scribble down a few key words describing the topic. With these hints, the conversation became easier to follow. "But these experiences," Zohra wrote to me, "though uncomfortable and sometimes stressful and frustrating, in a way, have made me happy, as they made me realize that I actually hear more than I thought I could."

David Wright points out that one problem with deafness is not so much "not hearing" as "not overhearing." Even when we aren't talking with people directly, we can often pick up the tone and words of their conversation, allowing us to gauge overall mood. Since Zohra depends upon both hearing and lip-reading to understand speech, she has a hard time overhearing. In 2017, Zohra and I enjoyed a visit to the aquarium in Toronto. It was a highly visual adventure, with glass cases full of spectacular fish and other creatures. People around us were exclaiming excitedly, but Zohra, perhaps preoccupied by the displays, did not overhear what they said.

Yet, on other occasions, Zohra suspects she is taking in more information from overhearing than she is aware. Once, in college, she was with her friends in the school's dining hall. The place was very noisy, so she didn't attempt to follow the conversation. Instead, she began thinking about a particular homework assignment. When she asked her friends if they had worked out the answer to homework problem five, they exclaimed, "Why, that's just what we've been talking about!"

In his memoir *Rebuilt*, Michael Chorost describes his frustration with understanding speech through a cochlear implant.[7] One day, while driving, he attempted to follow the voices on the car radio, but they all sounded like they were speaking "pseudo-English." So he began to think about other things. Then, quite suddenly and unexpectedly, he began hearing and understanding whole sentences! He found that if he concentrated intensely, the result was the same as if he paid no attention at all: he understood nothing. "You have to be calm, open, relaxed, alert. Poised at exactly the right mental place between idleness and tension."

Zohra had a similar experience around New Year's Day of 2018. She and her family were riding on a sightseeing tour bus, and at one point, when Zohra was relaxed, but not too relaxed, the words of the tour guide came through, loud and clear: "1926 . . . Our next stop is the Central Harbor Front . . . Lake Ontario . . . Together they make up 21% of the world's fresh water." What a contrast this ease of listening was to some of her efforts in her speech therapy sessions. There, the therapist encouraged her to concentrate harder. On some days, no matter how intensely Zohra concentrated, she still couldn't get the words. Listening harder just didn't seem to work.

David Wright also wrote about the delicate balance between tension and relaxation when it comes to lip-reading.[8] He loved going to pubs where, relaxed, he found it easy to understand and enjoy his friends. He had a deaf friend with an extraordinary ability to lip-read who could follow the speeches at a professional conference throughout the day. When Wright asked his friend for his secret, the man replied, "Relaxation."

All these experiences echo the theme of W. Timothy Gallwey's *The Inner Game of Tennis*, a book first published in 1974 and still popular today.[9] After years of teaching tennis, Gallwey realized that one of the greatest impediments to better performance was not a lack of speed or coordination but the way his students talked to themselves during a game—"I'm swinging the racket too late, my serve stinks!" We've all been frustrated by similar situations. We may be playing a sport well, moving easily and fluidly, until we think about our performance. Then our game falters. We may be playing the piano beautifully, just letting our fingers do the work, until we start telling ourselves how to play, thus breaking the flow. When Zohra tensed up during a speech therapy session in an attempt to follow speech better, her attention was now diverted between understanding the words and telling herself how she was doing. What she and all of us need to do, as Gallwey emphasized, is to engage in the activity without self-judgment and self-consciousness. At some level, we've always known this. Consider the common expressions "we need to get out of our own way" or "lose ourselves" in the activity.

With vision, when people tense up, they lose their view of the visual periphery, concentrating more on what is front and center.[10] This reduction in the area they see may focus their attention on what is most important at the time, but it does mean losing a view of the "big picture" or the context of the scene. Perhaps the same phenomenon happens with hearing. If we try to catch every word, we may lose the overall context, and without context, it's harder to follow what the conversation is about. Wright's friend, the expert lip-reader, is a case in point. Lip-reading is a poor substitute for hearing speech because only about 30 percent of speech movements can be seen.[11] As a result, the lip-reader must make educated guesses based on the context of the dialog. Absorbing the broad context is where Wright's friend, who remains open, receptive, and relaxed, even at a busy conference, may excel. We may take in much of the context subconsciously, and access to this information may be disrupted by too much tension and self-evaluation, making it harder to be "in the zone" in order to move accurately, see the big picture, or follow speech with fluidity and ease.

WITH THE COCHLEAR IMPLANT, ZOHRA'S OWN SPEECH IMPROVED considerably. Prior to the implant, very few people could understand her. At that time, she had to think hard about the movements of her tongue and the shape of her mouth in order to enunciate words clearly. After her implant, a transition occurred, one she was not aware of until, one day, she ordered a latte at Starbucks. The coffee shop was so noisy that she couldn't hear herself speak, causing her to hesitate before giving her order. That's when she realized how much she now depended upon hearing her own voice in order to talk clearly.

Because she had been reading all her life and learned words through reading instead of hearing, she used to pronounce some words the way they are read. For example, she used to say "cookumber" for "cucumber" and "cast-le" for "castle." By the time I met Zohra, ten years after she received the cochlear implant, her speech was almost always clear and intelligible, though it did sound a little unusual. Her voice is high and breathy, with a short-short-short-long rhythm. Some people, hearing the timbre and cadence of her speech and unaware of her deafness, have asked her if her first language was French!

– chapter 15 –

Talking to Herself

When I was cross-eyed, I was very cognizant of what was immediately in front of me but had little awareness of the visual periphery. I focused on details, not only when looking out upon a scene but in thinking about a problem. My husband, Dan, has the opposite approach. He has excellent distance vision and a great awareness of the visual periphery. But he hardly notices what's right in front of him. While I take care of the day-to-day tasks, Dan is always planning our next vacation or place to live. This makes me wonder, Does our perceptual style affect our way of thinking, and does our thinking style affect our perception?

While observing young children, the great psychologist Lev Vygotsky pondered the relationship between thinking and language.[1] When we are very young, we work through problems by talking out loud. By the time we reach five years old, this sort of monologue has started to turn inward. It becomes "inner speech," private and in our heads. Because inner speech is speech for ourselves, we use shortcuts. We may leave out, for example, the subject of our sentences. Eventually, the speech may become so abbreviated, it's hardly speech at all. At this point, we may, as Vygotsky described, be "thinking in pure meaning."

But Vygotsky's description does not apply to a child who is deaf. As a baby, Zohra didn't babble, and as a toddler, she didn't talk to herself. Although she learned to read when very young, she didn't think by seeing, in her mind's eye, the words of her thoughts scroll by. Even today, she rarely hears words in her head.

Mental imagery and memory play important roles in our thinking, and our imagery and memory are shaped by the way we perceive. Although Zohra's visual imagery and memory are very strong, her auditory imagery and memory are not. If Zohra hears a phrase repeated over and over again, she may remember it, but only for a little while. To train Zohra's auditory memory, Najma would cover her mouth and say a sentence that Zohra would repeat. Then Najma would say two sentences for Zohra to repeat, then three sentences, and so forth.

Zohra has no auditory memories from before her cochlear implant, and her sharpest auditory memories now are of emotionally charged events. Once when she was crossing a dangerous street, Najma yelled, "Stop!" Zohra can still imagine that event and hear the word "stop" in her head. She can recall the sound of her sister's baby crying when they talked on the phone. Indeed, she remembers conversations over the phone better than conversations face-to-face, perhaps because, with the phone, she is forced to depend entirely on her hearing. Liam, in contrast, because of poor vision throughout childhood, has excellent auditory but poor visual imagery and memory. He once told me, "I don't really have a mental picture of anything; as soon as I look away from something or at something new, it's gone." But, like Najma with Zohra, he has devised strategies for working on visual imagery and memory, for example, by drawing pictures of what he has recently seen.

When Zohra recalls her friends talking to her, she imagines their lips moving. When she reads their texts, she also sees their faces and lips. When I think of my friends talking and when I read their texts to me, I have an auditory image. I hear their words in my head and hear my responses too. Zohra's experience also contrasts with that of the poet David Wright, who lost his hearing at the age of seven due to scarlet fever.[2] Throughout life, Wright retained an "inner ear." As he wrote his poems, he still heard their words and rhythm in his head. With his inner ear, he heard the sounds that accompany movement. "The visible," he writes, "appears audible."

Zohra's congenital deafness, her particular way of encountering the world from birth through childhood, influenced the way she thinks. She never heard words as a child, so she doesn't think in words today. How then did she recall my lectures and do so well on my exams? This was the topic of our conversation back in 2011, while we enjoyed lunch together at a Chinese restaurant in Boston. Zohra had graduated from Mount Holyoke College the year before and was working in an audiology lab at Massachusetts Eye and Ear.

Zohra explained that she always sat in the front of the classroom to have a good view of my face as I talked. My frequent use of the blackboard was a great help to her. The blackboard was like a "captioned summary" of what was going on in class. She consulted the textbook carefully for what she might have missed in lecture. With these strategies, she absorbed the content of my lectures, their message, not the words. Liam said the same thing about the way he remembers what he's read. He doesn't see, in his head, the words on the page (as I do) but remembers their import. All this sounds like Vygotsky's elusive "thinking in pure meaning."

Yet, as I consider my lectures, let's say while walking to work, I hear myself speaking in my head. It seems like I am thinking in whole sentences. When I really scrutinize how I think, however, I realize that mentally rehearsing my lectures and thinking in general are very different. Often my thoughts start with visual images that I then bring to sharper attention with words. And, while I do like to talk to myself, there are many times when I think purely in pictures. I remember once swimming while trying to imagine how I was going to pour something from a large bowl into a narrow-necked bottle. Gradually a picture of a funnel emerged in my mind. Another time (also while swimming!), I pictured how a slide rule works—how sliding one part of the rule against the other is the manual equivalent of adding logarithms, which is the same as multiplying numbers.

Art historian Rudolf Arnheim wrote beautifully about visual thinking.[3] He pointed out that visual images are often vague and impressionistic. If asked to think about an elephant, many people see a visual image of the animal in their mind's eye. But if they then tried to draw the image, they'd discover that the details are missing. Arnheim drew a parallel here to impressionistic paintings in which the artist uses a few brush strokes to represent a person or a tree. Though lacking in detail, such images convey a great deal of information and a sense of movement and dynamism. Arnheim describes mental imagery as hints and flashes, a pattern of visual forces, and these can be very abstract.

We think not only in visual images but spatially. Indeed, this may be our most fundamental way of thinking since we all had to figure out, as babies, how to move through space and manipulate objects.[4] As I watch my infant granddaughter play, I realize how many problems she is teaching herself to solve, even without language. Just the other day, she learned which way to orient a spoon to get it into a jar. By nesting one toy cup into another, she learns about size differences—a helpful concept, but only if we remember to use it. When I challenged students in my introductory biology class to rank-order various large molecules and cell structures according to size, they were initially hesitant until I suggested they imagine which structures fit inside others. As I described in Chapter 8, we make cognitive maps of the landscape around us, but we also use cognitive maps in more abstract ways.[5] This is obvious from our language. If I say, "I am close to my brother," this phrase may mean that my brother is physically near me or that my brother and I share common feelings. With the exception of astronauts in orbit, we all work against gravity. We defy gravity when rising up but give in to gravity when falling down. So the up direction is associated with positive feelings and the down direction with negative ones. An "uplifting" song may help when we're in a "depressed" mood.

Any particular thought or mental image may include a vast amount of information. To pass on that information to others, to let them know what we are thinking, we use language. Yet words are ambiguous. A given word, such as "bat" or "bark," can mean more than one thing. Their meaning, as well as the assignment of pronouns, is only grasped through context. We can't pass on all the information in our heads, or we'll drone on forever and lose our listener's attention. We use shortcuts and expect our listener to infer the missing information.[6] These powers of inference are particularly strong in Zohra, who, throughout childhood, followed conversation primarily by lip-reading.

Zohra was a student in my neurobiology class, which focused on the electrical activity of neurons and synapses. As I lectured, I encouraged the students to see, in their mind's eye, the movement of ions through tunnel-like channels in the nerve cell membranes. I drew lots of schematic pictures on the blackboard. Indeed, my lectures were very visual and spatial. So Zohra remembered the content of my lectures in part through visual images, symbols, spatial arrangements and transformations, and mental logic, but to write down answers on my exams, she had to translate her thoughts into words. Perhaps, this is why she told me that "language is for communication between people while thinking is for communication between you and your brain."

– chapter 16 –

Musical Notes

IN JANUARY 2017, ZOHRA SENT ME AN EMAIL INFORMING ME THAT she had moved to Toronto, Canada, after having spent the last five years in her native Tanzania. Toronto is a lot closer to my home than Tanzania, so I jumped at the opportunity to see her again. Most of Zohra's extended family now lived in Toronto too, and I was eager to meet them all. This was a family who had pooled their resources to pay for Zohra's cochlear implant. Back in 2010, for Zohra's graduation from Mount Holyoke, Najma and Zohra's grandparents, then in their eighties, had traveled from Tanzania to Massachusetts. This was no easy trip: they had to travel to Kilimanjaro, Amsterdam, Washington, DC, West Virginia, where Najma's brother lived, and finally to Massachusetts. Surely the support Zohra's received from Najma and her loving, extended family carried over to the warmth she felt for other people and helped her to succeed in a hearing world.

In Toronto, Zohra lived in an apartment with Najma and her grandparents, who were now very old and infirm. They had initially come to Toronto seeking better medical care for Zohra's grandfather. Zohra's grandmother was suffering from Alzheimer's disease and often seemed distant and unfocused, except when she spied Zohra. Then her eyes, her whole body, lit up. Najma used a set of flash cards, with pictures on one side and English words on the other, to quiz grandmother on her English. Najma simply never stops teaching.

Often the apartment filled up with other members of the family, Zohra's mother and father, aunt and uncle, all of whom lived nearby and came daily to help Najma with the grandparents. Zohra's younger brother Ali drove in from Ottawa. He and Zohra are very close. He often translated the family dinner conversations for her, especially when everyone spoke Gujarati. With the mix of languages in the apartment and the comings and goings of various family members, I was reminded of stories of my own extended family, as they had been two generations before me, when my grandparents first immigrated to the United States. As for Zohra, when the conversation and activities in the apartment got to be too much, she simply removed the headset of her cochlear implant and read her book or smartphone in blessed quiet. Unlike most hearing people, Zohra is comfortable with total silence.

When Zohra graduated from Mount Holyoke in 2010, she was eligible for one more year of US residency on her student visa and spent it in Boston doing audiology research at the Massachusetts Eye and Ear Infirmary. Her boss there encouraged her to apply to graduate school, but she wanted to work in a hospital, a clinical research company, or the government. So she returned to Moshi, where she attended medical school. Medical school in Tanzania takes five years, so she graduated with her medical degree and moved to Toronto about a year before my visit in October 2017. At that point, she was earning a postgraduate diploma in clinical research. Though she misses Africa, she plans to stay in Canada with her family.

Shortly after I arrived, Najma fed me a delicious casserole of elbow macaroni and vegetables. When I asked her how it was spiced, she said with ginger and garlic, adding, "Like all Indian food." Najma's grandfather left India for Zanzibar in the early twentieth century, and later the family moved to East Africa, but they still keep their Indian cuisine.

As I had with Liam, I wanted Zohra to take the lead during my visit, showing me the places and telling me the stories about her hearing that were important to her. After the meal, Zohra put on her head scarf, grabbed a pink-and-white bag, led me down to the apartment's garage, and took me out in her new blue car to a coffee shop. Zohra reminded me that she wouldn't talk while driving because she couldn't look at my face and the road at the same time. But listening to music didn't pose a problem, so she turned on the radio and tuned it to a station broadcasting songs with a strong beat. This surprised me because, back in 2010, Zohra had told me she rarely listened to music. At the end of my visit, when her brother drove me to the airport, Zohra rode along in the car and asked for the radio and music.

Deafness does not prevent a person from having a good sense of rhythm. Rhythm is built into us. It is in our heartbeats, breathing, brain waves, and movements. As a child, Zohra was great at jump roping. André Aciman, in a *New Yorker* article about his congenitally deaf mother, wrote that she had a gift for fast dancing.[1] David Wright, deafened in childhood, at first assumed that he couldn't dance. But then a girl persuaded him to try, and he found it easy to follow her rhythm.[2] Most impressive of all is Evelyn Glennie, the world-renowned percussionist, who began to lose her hearing at age two, becoming profoundly deaf ten years later.[3]

But appreciating music requires more than a sense of rhythm. A musical note is a periodic sound; that is, it is made up of a fundamental frequency and harmonics. Frequencies are measured in the number of sound waves per second, or hertz (Hz), and a higher frequency gives us the sensation of a higher pitch. Harmonics occur at frequencies that are multiples of the fundamental. Playing the note A 440 on the piano or violin will produce sound waves at frequencies of 440 Hz (the fundamental frequency) but also at 880 Hz (two times 440), 1,320 Hz (three times 440), and so on. We don't hear these higher frequencies or harmonics as separate sounds. We hear one tone coming from the piano or violin. Though the note, as played by the two instruments, sounds a little different, its pitch will be at the same fundamental frequency, in the above case at 440 Hz. We even have an area of the brain lateral to the primary auditory cortex where neurons are selective for different pitches. They respond to both a pure tone (just the fundamental frequency) or to a tone made up of the fundamental and various combinations of harmonics.[4]

Pure tones, sounds of only one frequency, are very rare in nature. They are also irritating; the sound of an old microwave beeping is close to a pure tone. Since we can hear pure tones, why don't we hear each component of a sound separately? Why, when we are listening to music, do we hear a note, its fundamental and harmonic frequencies, as one tone rather than as many separate pitches? While the answer is not completely known, our way of hearing makes ecological sense. A major role of hearing is to identify a source from its sound. Even sounds that are nonperiodic and nonmusical, like a shotgun blast, are made up of more than one sound wave frequency. If we heard each frequency as a distinct pitch, we wouldn't be able to recognize the source of the sound. We wouldn't know where the different sound waves belonged. As with vision, where we see the whole of an object before its parts, we simplify what we hear, combining the sound waves produced by one source into a single recognizable sound.

Why does the note sound different when played on the piano or violin if it is at the same pitch? Each instrument has its own timbre, a somewhat elusive term that suggests the piano-ness or violin-ness of a note, a sensation that is independent of the note's pitch or volume. The contribution of each harmonic to the overall sound of the note differs from instrument to instrument, and this plays a large role in our sensation of the instrument's timbre.

Cochlear implants do not provide the sensitivity to pitch and timbre that is provided by normal ears. Very few cochlear-implant users can distinguish two notes that are only a semitone apart, such as C and C-sharp, and even distinguishing notes half an octave apart can be difficult.[5] As a result, even a simple melody, such as "Doe a Deer," may be hard to follow. When Zohra and I listened together to a song with a strong beat, Zohra mentioned that she couldn't tell if the singer was a man or a woman. The performer was singing in the tenor range, the higher registers for a man, and Zohra did not have enough pitch sensitivity to distinguish his voice from a woman's low alto.

Training a person with a cochlear implant to appreciate music may improve his or her speech perception and vice versa.[6] Spoken vowels are periodic sounds. Like musical tones, they are made up of a fundamental frequency and harmonics. Since vowels make up all syllables and words, our voices have pitch. As our voice rises and falls, the fundamental frequency and all the harmonics of a given vowel rise and fall together, facilitating our ability to recognize and follow a given person's voice. It is possible that Zohra's increasing ability to understand speech improves her ability to hear music, which in turn enhances her speech comprehension.

For people who once heard through normal ears but now depend upon a cochlear implant, music may not sound as rich.[7] If it's difficult to hear the difference between one note and another, it may be hard to distinguish, for example, a song in the major versus minor key. So, some of the emotion and expression that is part of the music may be lost.

Those with the most musical experience before their hearing loss may fare best. Arlene Romoff grew up in a musical household and was a gifted pianist before her hearing began to fail in her late teens. The first music she enjoyed when she received a cochlear implant in her left ear was Herbie Mann's jazz flute.[8] A flute provides a relatively pure tone with fewer higher harmonics than other instruments, and this may be why she found it easy to hear. But over time and with repeated listenings, she could appreciate even more. Upon first hearing an unidentified piano piece on her CD player, she recognized it only as sounding like Chopin. But a week later, she could pick out its distinctive bass notes and identify it as "Raindrop Prelude," a piece she had played thirty-five years earlier. One week after that, she heard the melody too.

Zohra has no memory or experience of music as heard through normal ears. To her, music, especially songs with a strong beat and little background accompaniment, sound good. In Toronto, she loves to go with a friend or cousins to Christmas Market during the holiday season. There, she not only sees but also hears performances by singers, dancers, and carolers. "So joyful to hear Christmas carols and festive sounds and share the moment," Zohra writes. As she has said to me repeatedly, her cochlear implant helps her to connect with the wider world. Enjoying music with others provides one more connection.

– chapter 17 –

The Cocktail Party Problem

When we arrived at the coffee shop, I realized that Zohra had chosen a good place to have our conversation. It was quiet; there was no background music, and we were able to sit at a table far from other people. For Zohra, as for most people with cochlear implants, it is hard to follow speech in a noisy environment, such as a restaurant abuzz with conversation. This challenge was dubbed the "cocktail party problem" by Colin Cherry in 1953, and it's analogous to the visual problem of distinguishing an object from its background.[1] Most of us are able to isolate a friend's voice from other sounds by focusing on its qualities, including its pitch range and timbre, but because Zohra hears through an implant, she has reduced pitch and timbre sensitivity.[2] Scientific studies of the human auditory cortex indicate that the primary auditory cortex represents all speech sounds, while higher, nonprimary areas selectively represent the speaker we are attending to.[3] So poor development of higher auditory areas may compromise Zohra's ability to focus on one speaker in a crowd.

What's more, Zohra has a cochlear implant in only one ear. If she had implants in both ears, she might have an easier time hearing speech in noise.[4] This is what Arlene Romoff discovered after she received a second cochlear implant and developed binaural hearing.[5] She could separate sounds and voices and follow conversation around a noisy dining room table better than with only one implant. Try plugging one ear next time you are in a noisy restaurant, and you too will find it harder to follow a given person's speech. Two ears are much better than one for pinpointing sounds.

ABOUT TEN YEARS AGO, A STUDENT OF MINE NAMED PAULA BOUNDED into my office with her ponytail bouncing behind her. "You look happy. You're wearing your hair differently," I said to her. Paula is partially deaf and used to cover her ears with her hair to hide her hearing aids, but, on this day, her hair was pulled back, and I couldn't see any hearing aids at all. When I asked her what happened to them, Paula quickly reached up and removed a tiny object from inside one ear and then showed me how the parts of the aid that hung around her outer ear were transparent. When she put the hearing aid back in her ear, the external parts seemed to disappear. But the hearing aids had improved her life in more ways than just appearance. "I love technology," Paula said. "My hearing aids keep improving. For the first time, I can tell where a sound comes from. I don't have to do this," and then she opened her eyes wide and moved her head back and forth as if to scan the scene in what was obviously a very practiced move. "When I hear a sound, I don't have to look for it. I just know where the sound is coming from."

Paula's new hearing aids made it easier for her to (unconsciously) compare the input from the two ears because the difference in that input helps us to localize sound.[6] Sounds to our right, for example, reach the right ear before the left and with greater intensity. But even people with one ear can localize sounds to some degree. The shell-like shape of our external ear, the pinna, filters incoming sound differently for different frequencies and locations, helping us to know where sounds originate. You can get a sense of the role of the external ear by listening to a recording with earphones or earbuds. Under these conditions, the sound bypasses the external ears and seems to come not from the outside but from inside your head!

Under some circumstances, we may be able to use sounds and their locations to produce a mental picture of a whole landscape, as John Hull describes in an extraordinary passage in *Touching the Rock*.[7] He was blind, but standing by his front door and listening to a steady rainfall, he perceived the whole scene—the sound of the water dripping from a drainpipe on the house roof told him the drainpipe was to his left; the water sounded differently on a leafy shrub located a little further on. With the changing sound of the raindrops, he could even hear where the lawn, fence, and path were located, right down to the garden gate.

This is exactly what Arlene Romoff experienced after receiving a second cochlear implant and developing binaural hearing: "The brain actually lays out a sound landscape, like a 3-D site plan. It's as if I had a map with a key identifying the locations of the sounds surrounding me."[8] Before her second implant, sounds came out of the blue, but now she felt immersed and enveloped in this landscape of sounds. This experience made her happy in ways that hearing with a single cochlear implant never had. I was very struck by her descriptions because I had the same feelings of immersion and joy when I first could coordinate my eyes and see in 3D.[9]

Zohra has a cochlear implant in only one ear, and the implant mechanism bypasses the external ear entirely. Instead, sound is picked up by a microphone placed behind her ear and is transmitted to a receiver in the implant. So I was curious as to whether Zohra could localize sounds. As we sat down at the coffee shop, I asked her about her sound-localization abilities. If a friend walked into the shop and called her name, would she know where to look? Zohra said no and moved her head all around to indicate how she would locate the sound. Then she said something really startling—that even the concept of a sound having a location is foreign to her. Unless she can see their source, sounds are completely disembodied and will remain that way unless she receives a second cochlear implant. As we walked by her apartment the next day, someone was trimming a hedge nearby and off to our left. Because of the high hedge, we couldn't see the instrument, but we sure could hear it. Yet Zohra could not locate where the sound was coming from.

Newborns will turn their eyes in the direction of sounds.[10] So our ability to localize sound is probably present at birth, although it improves considerably as we develop. This makes ecological sense. We can hear but cannot see around corners, so we can use sounds to locate things we cannot see. When we hear a sound, we turn our heads so that the sound falls equally on our two ears. Now its source is located straight ahead, the location where we see it best. Identification and localization are two different but important functions for all of our sensory modalities. Like with vision, where there are separate anatomical and physiological pathways for the identification and location of visual stimuli, there may be separate "what" and "where" pathways for the identification and location of sound sources.[11] With only one cochlear implant, Zohra's ability to locate and orient to sounds was lacking. No wonder the first sounds she heard through her implant were frightening and she has trouble following a friend's talk in a noisy environment.

But there was one strategy that Zohra could use to help locate sounds. When she first returned to Moshi, six months after receiving her cochlear implant, Zohra discovered something new. Her bedroom window faced a street, and at night, when all else was quiet, Zohra picked up the sounds of cars approaching her window. She could hear how the noise of their motors gradually increased as the cars approached and then decreased as they passed by. This was an epiphany. She hadn't expected things to sound louder the closer they came. Liam made an analogous discovery when he first saw a ball coming toward him while playing catch. Objects sound louder and loom larger when they approach, but for Zohra and Liam, it took experience to find that out.

– chapter 18 –

Zohra Damji, MD

FIGURE 18.1. Zohra in Toronto, 2019.

AT THE COFFEE SHOP, ZOHRA WAS MOST ANXIOUS TO TELL ME about medical school. The first two years involved classroom work. That was easy, she said, but added that her handwritten notes from class were very messy. Zohra has beautiful penmanship, but in class she couldn't read lips and take notes at the same time, so she wrote at a sprint. Starting in the third year, she began to observe and assist in the operating room, where everyone wore surgical masks that covered their mouths. This made it hard for Zohra to follow conversation.

Outside the operating room, Zohra faced an additional challenge. Most of the patients spoke Swahili, a language Zohra didn't understand. In her third medical school year, Zohra knew that she, like all her classmates, would be sent to a remote area to take patient histories and practice medicine. So she hired a tutor during the summer before her third medical school year and took lessons from mid-July to October. Her first interview with a patient didn't go well. She didn't understand anything, not because of her deafness but because of the Swahili. To get through her patient interviews, Zohra learned the most common questions and replies. At this point in the conversation, Zohra turned over the cloth bag she had brought with her and shook out its contents. Out spilled a ton of pink cards the thickness of index cards but cut to different sizes. They were handmade flash cards with Swahili words and phrases on one side and English translations on the other. For example,

Sijisikii vizuri—"I don't feel well."
Nimechoka—"I feel tired."

Zohra used these cards for a crash course in Swahili. But learning the words was not enough, as Zohra illustrated by writing two English sentences on a napkin:

"Do you have fever?"
"You do have fever."

She pointed out that you can tell the difference between the question and the statement by the word order. But this is not so in Swahili. She added to the napkin the Swahili translations:

Una homa?
Una homa.

To tell the difference between these two phrases you need to sense the upward inflection in speech, and that was hard for Zohra to hear.

The more Zohra interviewed patients, the more she could anticipate their replies. She found she could structure the interviews. She asked the questions; the patients answered. Psychiatry was the most difficult, so, if possible, she interviewed her psychiatric patients in English. If the patient's conversation sounded strange, she wasn't sure if the patient was hallucinating or she had heard their tale incorrectly. Pediatrics, on the other hand, was easier. Questions included, "Does the child have a fever?"

Had Zohra stayed in Tanzania, she most likely would have become an obstetrician-gynecologist, although she also considered "deaf-friendly" specialties, including ophthalmology, pathology, radiology, clinical research, and public health. In Canada, she is not yet sure how she will use her medical education except that it will be in a way to help others. What she loved most about medical school occurred during her third year when she was sent to a remote sugar plantation. She remembers stepping off the bus in an area of open grasslands filled with the smell of fermenting sugar and cane molasses. There was a ninety-bed hospital there that served people from the plantation and surrounding communities. Zohra was filled with trepidation. Could she, a deaf person, relate to her patients and become a good doctor?

Like Liam, Zohra solved her problems one by one. She ordered a special stethoscope that amplified heart and lung sounds and provided a visual display. During patient rounds, she carried a special blood-pressure device, digital and lightweight, which fit into her pocket. But above all she practiced her Swahili.

Zohra is enormously proud of learning Swahili. She had gotten an A in French class in high school, but that was classroom French, not something she needed to use every day for her job. And acquiring Swahili gave her far more than just a practical skill. "There is a deep sense of joy when you are able to talk to your patients, connecting with patients, learning their language, and establishing rapport with them," Zohra wrote to me. With her cochlear implant, Zohra taught herself to hear, reshaped her whole perceptual world, and, in doing so, was able to enter the world and lives of her patients. After her first two years of medical school, the dean confessed that, if he had initially known that Zohra was deaf, he would not have admitted her. According to the chief of the hospital, Zohra was the first deaf doctor in Tanzania.

Conclusion
Athletes of Perception

What we hope ever to do with ease,
we must first learn to do with diligence.

—Samuel Johnson, quoted in J. Boswell, *The Life of Samuel Johnson* (New York: Penguin Random House, 2008).

The one hundred and forty thousand transistors in
my skull give me sound, but they cannot make me *listen.*

—Michael Chorost, *Rebuilt: How Becoming Part Computer Made Me More Human* (New York: Houghton Mifflin, 2005), 183.

When, through optometric vision therapy, I first learned to coordinate my eyes, I began to see in 3D for the first time.[1] Ordinary scenes looked extraordinary. I could see the empty space between leaves on trees or between falling snowflakes. If I looked in the mirror, I was astonished to see my reflection not on the surface of the glass but behind the mirror in the reflected space. I was seeing not only in a new way but in a way that I couldn't have previously imagined. Nothing in my former visual experience prepared me for what it was like to see the space between.

Dr. Oliver Sacks was the first person, outside family and friends, to whom I told my story.[2] Since he had written about Virgil, a man blinded since infancy who gained but then rejected sight as an adult, Dr. Sacks was particularly curious about my reaction to stereovision. Did my new views ever confuse or disturb me? Sometimes, I told him. After seeing in stereo, I developed a fear of heights, a feeling brought home to me while hiking along a cliff face. But, mostly, my new sights filled me with a sense of wonder and a childlike glee.

My new stereo views brought me more joy than confusion because they reinforced and expanded my overall worldview. I could see the volume of space between things that, in the past, I could only infer were there. The world appeared to expand and inflate, but objects were still recognizable and organized in the landscape in the same depth order that I had always seen them in. I did not have to restructure my whole perceptual world. Everything made sense.

The same was not true for Liam and Zohra. Learning to see and hear required a radical change in the way they perceived the world and an exhaustive search for meaning. Where we see objects, Liam saw the lines and color patches that make up objects. For Zohra, all sounds—voices, a motor, the rain—initially merged into one confusing mix.

The great psychologist William James once speculated that an infant was born into a world of "blooming, buzzing confusion."[3] James's description reflects the chaotic and frightening world that Liam and Zohra found themselves in, but it almost certainly doesn't reflect the experience of a newborn baby. What infants need most from their senses is the ability to perceive and interact with their caretakers. Within days, they can recognize their mother's voice and face, and within the first six months, they become particularly sensitive to the sounds of their native language and the kinds of faces they normally see.[4] As little babies, they do not need to use their senses to cross a busy street, read print, or understand a voice on the telephone. Their perceptions match their needs, giving them information about the most important sights and sounds in their environment.

This was not the case for Liam and Zohra. Before their surgeries, they already had substantial knowledge of the world, but after the operations, they were bombarded with one meaningless sensation after another. To cope, they had to recognize that their new sensations belonged to familiar categories of living and nonliving things. Lumping things into categories makes the task of identification much easier.

Sometimes, when I walk to work along the same old route that I have taken for more than twenty years, I remind myself that each glance brings something new. I may leave the house at the same time each day, but the sun's position in the sky is always different. I walk by trees rooted to their spots, but their leaves grow, change colors, and fall off. I pass people I have never seen before, but I instantly recognize them as people and can even tell their gender, age, and often mood. A new bird may have recently flown into the neighborhood, but I immediately recognize its species by its call. Although I am seeing or hearing something new, I am hardly aware of its uniqueness because it fits into my overall world scheme. Trees, birds, people, and cars—they all fit into the categories of what I expect to see.

All members of a given category share common properties that we extract despite differences between the members. Discovering and picking out the relevant patterns results in perceptual learning. The patterns we extract may be represented in the brain by networks of interconnected neurons.[5] It is the network, not an individual neuron, that is responsible for the pattern, and a given neuron can be a member of many networks. Pattern recognition for sights and sounds requires more than raw sensory input and more than the primary visual and auditory cortices, respectively. Networks involving both lower and higher sensory areas are involved. As Liam and Zohra took in new sights and sounds, they had to classify them as belonging to familiar categories, categories they knew from their other senses.

When I first began to see in 3D, I joked with friends that my new vision stimulated my whole brain. I felt more alert, almost too sensitive, not only to sights but to sounds and other sensations as well. I began to listen more attentively to music and play the piano regularly. While reading Elkhonon Goldberg's book *Creativity*, I realized these feelings may have revealed a deeper truth.[6] Goldberg suggests that the left hemisphere of the brain is particularly important for storing patterns and categories. When, however, we see something unrecognizable, something that doesn't fit into the patterns stored in the left hemisphere, activity increases in the right hemisphere. While the left hemisphere deals largely with established patterns, the right hemisphere deals to a greater extent with novelty and uniqueness.[7] At first, I wondered how this dichotomy worked for vision since the right brain processes visual input from the left visual field and the left brain from the right. When we see something curious, however, we look directly at it so that it is positioned in the center of our visual field, activating visual neurons on both sides of the brain. Could my new views, so novel to me, have enhanced activation of my right brain, contributing to the feeling that my whole brain woke up as my vision transformed? For Liam and Zohra, their right hemispheres may have been continually stimulated by novel stimuli. As they began to recognize what they saw or heard, new sensory networks may have developed, especially in the left hemisphere. So learning to see and hear involved changes throughout both hemispheres of the brain.

Adolescence is a time of increased risk taking and novelty seeking, and this is the stage in life when Liam and Zohra first experienced dramatic sensory changes.[8] At this age, they may have been better able than adults to handle the onslaught of novel sensations. And as adolescents, their personal identities and societal roles were still developing. An individual who gains vision as an adult may transform from being a competent, contributing blind person into a less competent, less functional sighted one. This was certainly true for SB, the man who gained vision following a corneal transplant at age fifty-two.[9] Richard Gregory and Jean Wallace described how he was a competent tradesman before gaining sight. Extroverted and confident, he was proud of his accomplishments, but when he began to see, he found his achievements to be paltry, not the equal of what a sighted person could do. He grew increasingly depressed and withdrawn. Gregory and Wallace wrote, "He certainly relied a great deal on vision, but we formed the impression that this very reliance cost him his self-respect." At age fifteen, Liam was not expected to hold a job, be a breadwinner, and take care of others. He was still at an age where he was supposed to learn everyday skills. He did not yet have a career. This gave him more freedom and more time than a fully mature adult to learn, fail, try again, explore, and incorporate his new ways of seeing into his sense of self.

Nevertheless, Liam and Zohra had to adapt to a society that assumes that people see and hear. If you had met the two as young children, you might never have guessed just how smart, competent, and resourceful they are. Liam received so little information from looking at a face that he didn't make eye contact, and he was too shy to talk to people he didn't know. Zohra may have observed you carefully, but her speech was hard to understand. Both were able to adapt to our seeing and hearing society, succeed academically in regular schools, and adjust to their new sense because of support from family, doctors, and therapists and, most especially, the unfailing devotion of Cindy, Liam's mom, and Najma, Zohra's aunt. Not only did Cindy and Najma advocate for them and provide guidance and training, but they taught Liam and Zohra the discipline they needed to learn to see and hear and gave them the confidence to do so.

Today, the hard-earned skills that Liam and Zohra acquired at home and in school pay dividends. Given Liam's poor visual acuity and the need to enlarge print to enormous size, it would have been easiest for him to avoid print reading and just learn Braille. But he learned both—to read print in regular school and to read Braille, thanks to his mother's insistence, in special classes. Many people with childhood blindness who recover sight as adults learn to recognize printed letters but not to read print, even if they are proficient with Braille. Liam's ability to see small print and then to read and understand it has been of enormous benefit.

Most children develop language by hearing others talk, so a child born deaf is at risk of never acquiring language. But Najma taught Zohra English through reading. Before or by the age of three, Zohra had language, an enormous achievement and benefit; yet she could not easily communicate with others. Despite all the hours of talk training that Najma provided, Zohra's speech was barely intelligible. But when Zohra received her cochlear implant and could hear herself speak, she had all the English she needed to learn to speak clearly.

Adolescence may be a time for exploration, but it is also the age when we may be most self-conscious.[10] As Liam and Zohra saw or heard better, they became acutely aware of how much better other people could see or hear them. With her cochlear implant, Zohra was embarrassed to discover the loud chewing noises she made when she ate or the shuffling sounds she produced when she walked, two habits she quickly abandoned. In a large airport, Liam got lost and wandered into restricted areas. This was embarrassing. So he took out his cane, hoping that people would assume he was blind, not stupid, and would be more likely to help.

Zohra isn't afraid to ask others about new sounds, and she possesses such warmth, calmness, and contentedness that she is never without close friends. While in college, knowing that she wanted a clinical career, she joined the Association of Medical Professionals with Hearing Losses. At their conference, she met Dr. Samuel Atcherson, a PhD researcher and audiologist who also wears a cochlear implant. He is a great inspiration to her, and she spent part of her summers between college years doing auditory research in his lab. Liam doesn't judge himself harshly but sees his challenges for what they are. As his vision improved, his desire to be part of a group and to help others increased. He feels a strong bond with other people with low vision or albinism and has joined organizations, such as the National Organization for Albinism and Hypopigmentation (NOAH), that support and advocate for them. The better Liam and Zohra can see and hear, the more they can communicate and share their lives, thoughts, and skills with others. Their stories highlight the need we all have to be part of human society and how our perceptual abilities influence and are influenced by our ability to do so.

Despite her hearing loss, Zohra has great confidence in her own perceptions, as became obvious during an ethics class that she took in Toronto for her postgraduate diploma. Zohra was the fifth person to enter the classroom that day, which, unbeknownst to her, made her the subject of a classic psychology demonstration, the Solomon Asch conformity experiment.[11] The four students who had arrived in the class before her had just been prepped on how to behave. To carry out the experiment, the professor showed a PowerPoint slide of two cards (Figure C.1). Although the line on the left card was the same length as the rightmost line on the right card, the students were instructed, upon Zohra's return, to tell her that the middle line on the right card matched the single line on the left card. Zohra, the professor predicted, would capitulate to groupthink and declare the line on the left card as equivalent to the right card's middle line. The four students enthusiastically embraced their acting roles, pretending to think hard before claiming that the left card line was the same length as the middle line on the right card. But Zohra didn't bow to peer pressure. She maintained that the line on the left card was the same length as the rightmost line on the right card. The teacher was taken aback, mumbling that Zohra was "very brave." But having grown up deaf, Zohra must put great trust in her interpretation of what she sees, and, most important, she isn't afraid to be different.

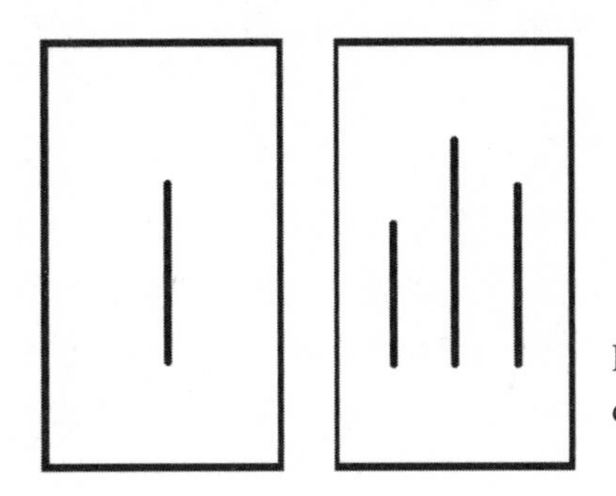

FIGURE C.1. Asch conformity test.

When Liam received his intraocular lenses, he wanted to remake himself as a visual person. He folded up his cane, put it in his backpack, and walked along the streets guided primarily by vision. This worked well enough when he frequented only familiar places around school and home and, later, when he worked mostly at night and was not bothered by the sun's glare. But on a rainy night in May 2015, he took a bad spill on his bicycle and began to ride less often, particularly if he had to ride on busy city streets. Over time he learned that he was most comfortable walking with both his vision and his cane. His vision might suggest a set of stairs just ahead, a situation he could confirm with his cane.

Liam struggled with this decision because others questioned him, pointing out that his visual acuity should be good enough to walk without a cane. Indeed, "low vision" is legally defined as best-corrected acuity between 20/70 and 20/200 in the better-seeing eye. Liam's acuity usually measured around 20/60. But his visual acuity was evaluated in a doctor's office, in evenly lit surroundings, while he was sitting still and reading an unmoving eye chart. It was an entirely different situation to see well on a busy, sun-drenched city street, full of glare and shadows, and with people and cars moving in all directions. Liam learned that visual acuity, which measures the size of the smallest detail you can see at a given distance, doesn't predict other essential visual processing skills, especially how well you can assimilate and understand visual information. His situation is very different from those who suffer vision loss later in life. They can maximize their remaining sight using previously learned visual skills, although they must acquire, for the first time, the tools of the blind. Liam, on the other hand, may not process visual information quickly, but he is experienced in using a cane and reading Braille.

When Liam moved to St. Louis after college, he joined sports teams for the visually impaired and NOAH. Through these groups, he learned about bioptics, small lenses that fit over regular glasses and magnify from two to six times. He also consulted Missouri's Rehabilitation Services for the Blind, which provided him with bioptics and a refresher course on cane navigation. He uses bioptics mounted on his glasses for reading the computer screen at work and menus posted on restaurant walls. Tinted contact lenses help with glare, but if the light is still too bright for him to see his phone screen, he switches to the Braille reader paired to his smartphone. Zohra, too, has learned to exploit all her abilities, using both hearing and lip-reading to understand speech. While they have different sensory abilities than most people, Liam and Zohra are making the same choices that any of us make—prioritizing and tuning their particular perceptual skills for the lives they want to lead.

Liam was just fifteen when he received his intraocular lenses and Zohra twelve when she obtained a cochlear implant. At these ages, our brains are not fully mature.[12] This conclusion is based in part on observations of overall structural changes in the brain that occur across the lifespan. As we mature, we develop long-range connections between far-flung brain networks. Axons that connect these networks must conduct nerve impulses rapidly, and they do so because they are covered by a fatty sheath called myelin. The process of myelination, which begins *in utero*, is not complete until early adulthood. Indeed, for some areas of the brain, such as the frontal cortex, development may not be complete until the forties! So Liam and Zohra gained a new sense at a time when their brains were not fully developed, perhaps making it easier for them to reshape their brain circuitry.

However, not all parts of the brain develop at the same rate. The first brain areas to mature are those involved in sensory processing and movement.[13] In the first years of life, our sensory systems tune themselves to the stimuli most commonly encountered. What happens, however, to the auditory areas of the brain if you are deaf and the visual areas if you are blind? Brain imaging experiments reveal that these areas don't disappear but may be taken over, in part, by the remaining senses. Thus, people profoundly deaf from birth recruit the unused auditory cortex for seeing.[14] Blind people use neurons in the visual cortex for monitoring the direction of sounds and for reading Braille.[15] Indeed, even sighted people, if blindfolded for five days and given intensive Braille lessons, will recruit visual cortical neurons when reading Braille, though this recruitment will end shortly after they're allowed to see again.[16] It is very likely then that neurons and circuits in Liam's visual cortex and Zohra's auditory cortex were co-opted to some extent by their other senses in early childhood, and these changes may influence their ability to use their sight and hearing today.

Could there be some brain logic to this co-option by the remaining senses? Our experience of the world is multisensory. A car passes by, and we both see and hear its movement. Indeed, visual areas involved in movement perception may be reallocated in a blind person for sensing movement through touch or hearing.[17] Auditory areas involved in understanding oral speech may be used in a deaf person for the visual task of understanding Sign.[18] What happens to these areas if the lost sense is restored? Can the newly sighted person, for example, repurpose the visual cortex for sight? Michael May, who was blinded at age three and regained sight at age forty-six, has good visual motion sensitivity. Brain imaging experiments reveal that the visual motion area, MT+, "lights up" in May, as it does in normally sighted people, when he sees movement.[19] This same area is activated in May, but not in normally sighted people, when he hears something move by. So MT+, a visual area in most people, has become multisensory in May; that is, it serves both vision and hearing. Nevertheless, this change doesn't seem to have compromised May's ability to see motion.

WE ALL DEVELOP SENSORY HABITS WHEN YOUNG. OFTEN WE ARE not even aware of them, and these habits may help or hinder adaptation to novel sensory experiences later in life. For me to learn to see in 3D, I had to abandon my old cross-eyed way of moving my eyes (looking with one eye and turning in and suppressing the other) and develop new patterns (pointing the two eyes simultaneously at the same place in space).[20] SB may have recovered sight at age fifty-two, but this didn't change the way he took in the world. He didn't habitually scan his surroundings with his eyes, as normally sighted people do, but looked at things only when they were pointed out to him.[21] If Liam wants to get better at recognizing faces and facial expressions, he must develop the habit of looking at people's eyes and faces. In 2018, he undertook vision therapy with a developmental optometrist. One exercise revealed that he sees with a very narrow point of view. When he attempted to read a letter chart located directly ahead while walking around chairs arranged in a figure eight, he kept bumping into the chairs. Since he must focus intensely on what is in front of him, he pays less attention to the visual periphery despite having full peripheral vision. When the letter chart in the exercise was replaced by a simple light, Liam could keep his gaze on the light but still see and avoid the chairs. Exercises, such as this one, may help him readjust the balance between his central and peripheral vision.

When Michael Chorost received his cochlear implant in 2001, he was given no auditory training whatsoever.[22] Although he received substantial support from his audiologist in terms of adjusting or mapping his implant, there was no institutional structure, no set of resources or step-by-step plan, to teach him how to hear again. In his memoir *Rebuilt*, he writes, "Fifty thousand dollars had just been spent on me, but it wasn't the end of my reconstruction. It was just the beginning. But with all that our civilization knows about human performance, you'd think I would have coaches, training programs, multimedia CD-ROMs, the works. Wrong. No formal training of any kind was offered to me during those tumultuous and difficult months after activation."

In 1997, Mark Ross, professor emeritus of audiology at the University of Connecticut, published an article in the *Journal of the Academy of Rehabilitative Audiology*, in which he reflected on his own experiences with aural rehabilitation.[23] In January 1952, while in the military, he was sent to Walter Reed Army Medical Center to be fitted with hearing aids. He joined a program that had been developed shortly after World War II, at a time when the military faced a large number of servicemen who had lost their hearing during combat. Ross spent about eight weeks in the hospital, where he received a comprehensive evaluation of his hearing, individual and group therapy, classroom instruction, and tests and retests of his hearing aids. The classes, which focused on speechreading, were creative and often playful and included visual perceptual and memory exercises, skits for identifying verbal and non-verbal messages, and training that progressed from the discrimination of broad speech features to recognition of smaller and smaller acoustic differences. For some servicemen, occupational and vocational counseling was also provided. As Ross thought back to this experience and its benefits, he described it as an "audiological Camelot."

But in the latter part of the twentieth century, rehabilitation programs, including the one at Walter Reed, disappeared.[24] For audiologists, much more emphasis was placed on the technical aspects of their profession, such as diagnostic tests and hearing devices, than on rehabilitation training. While therapies were often provided to help an adult adjust to deafness, fewer therapies were developed to help an adult take advantage of a gain in hearing made possible through better hearing aids or cochlear implants. This shift away from rehabilitation was influenced to a large extent by the scientific wisdom of the day, which maintained that sensory abilities and language develop only during critical or sensitive periods in early childhood.[25] Beyond the first eight years or so after birth, there was little possibility of further growth. So the ability of adult patients to benefit from a better hearing device was limited by what skills they had developed as infants and young children. Training for adults, it was concluded, was a waste of time. What a contrast this mind-set presents to the training of adult athletes, in which every muscle, every movement, every emotion is measured, analyzed, and optimized. There is little doubt that these athletes, though developmentally mature, can improve.

To hear again, Chorost had to become "an athlete of perception": "I had to learn to glide over the sound-stream like a skier over bumpy snow, to assemble meanings out of phonemes like a juggler keeping ten balls in the air."[26] He devised his own therapy exercises, for example, listening to children's books on tape while reading along. Beverly Biderman came up with the same strategy, and Zohra watched and listened to videos while reading the closed captions.[27] After receiving her cochlear implant in 2000, Zohra did receive some formal speech therapy, because, unlike Chorost, she was still considered a child. But the bulk of her therapy, both before and after her implant, was provided by Najma at home. Although Liam received therapy as a child to cope with increasing blindness, he received no training after receiving his IOLs. Yet, as early as 1913, F. Moreau, a surgeon who performed a sight-recovery operation on an eight-year-old boy, wrote, "It would be an error to suppose that a patient whose sight has been restored to him by surgical intervention can thereafter see the external world . . . The operation itself has no more value than that of preparing the eyes to see; education is the most important factor . . . To give back his sight to a congenitally blind patient is more the work of an educationist than that of a surgeon."[28]

Research, largely performed in the twenty-first century, has revealed that the adult brain is far more plastic than initially believed, and new neurobiological mechanisms for this plasticity are continually being uncovered.[29] While an adult brain is not as malleable as a young one, more and more clinical and scientific papers support the benefits of training, even in older adults, who suffer vision and hearing losses.[30] Yet many of these papers confine their reports to the results of clinical tests—for example, how many lines on the eye chart a patient can read or how well a person can distinguish words in a noisy environment. While clinical tests evaluate isolated skills, they give no hint of what it feels like to see or hear in a new way or how these new abilities impact daily life. A patient's personal experience is considered "subjective" or "anecdotal," descriptions that convey the kiss of death in scientific circles.

As a result, only passing mention is made in many clinical studies of the patient's emotional reaction to the training. Yet, for training to be successful, a patient's feelings and attitude are critical. No matter how good the results of the training may be, if the patient does not find them relevant or meaningful to daily life, or if the training is tedious, boring, and unpleasant, then he or she will not follow through. As Arlene Romoff, who learned to hear again with cochlear implants, wrote, "The ultimate aim is not finding perfection in the testing booth. It is about behavior, not test scores—functioning at ease in the hearing world."[31] Learning to see or hear as an adult or older child is more than a mechanical retuning of the eyes and ears or the development of a collection of skills. The acquisition of a new sense uproots your whole perceptual world. Novel sensations and skills must be integrated and assimilated into your personal worldview and way of being.

Laboratory experiments reveal that learning and changes in brain wiring in the auditory and visual cortices are facilitated by the liberation of neuromodulators from other areas in the brain, such as the brainstem and basal forebrain, and this release is triggered by novelty and anticipation of a reward.[32] So if patients expect a positive experience from their training, they, by the liberation of neuromodulators in their brain, are enhancing their own plasticity. Such studies support the important role of a patient's attitude to his or her therapy. My friend's daughter, Kaitlin, was born with cerebral palsy. By the time she turned ten, she was putting on weight and struggled to walk with crutches. Physical therapy was not working. So my friend hired a physical therapist who was also a personal trainer who made house calls. This changed everything. In Kaitlin's mind, people with injuries and disabilities see physical therapists at the hospital, but the rich and famous have personal trainers who come to their homes. Kaitlin worked hard with her trainer, stuck to her diet, slimmed down, and progressed to walking one mile per day.

SB, the blind patient described by Gregory and Wallace, initially experienced excitement and delight when he first could see.[33] But this delight turned to despair when he realized that he would never see as a normally sighted person. Seeing became a burden, and less than two years after his sight was restored, he grew terribly ill and died. In her memoir *Wired for Sound*, Beverly Biderman too wrote about the despair she felt several months after receiving her cochlear implant.[34] She couldn't trust what she heard, didn't like many of the sounds, and couldn't relax. Deeply disappointed in her new life, yet unable to go back to the old one, she lost the will to continue.

What helped turn things around for Biderman was auditory training. Six months after she received her implant, her clinic began an auditory rehabilitation program. Working with her therapist, Biderman learned to distinguish words that sound similar, such as "tab," "tap," and "tack." Her therapist would announce a topic and then present sentences one by one on that topic. "If I squeezed my eyes shut and concentrated hard," Biderman wrote, "I could understand her and repeat whole sentences quite successfully. I was drained emotionally and physically after these sessions, but exhilarated and inspired too." She began to relax and enjoy what she heard. She felt a sense of control over and active engagement in her own progress. Her experience is not unique. Researchers at Advanced Bionics, a leading manufacturer of cochlear implants, now recognize the importance of sensory training for adults. They offer a program called SoundSuccess that lets users listen to various sentences and take quizzes to evaluate how well they hear them.

When I undertook vision therapy for my crossed eyes, I also experienced a feeling of empowerment.[35] I grew increasingly aware of how I used and moved my eyes, learned to troubleshoot problems, and no longer saw myself as a victim of a sensory deficit that had hounded me since early childhood. Just as Biderman described, people seemed kinder, and the world became a less hostile place.

The great writer and thinker Samuel Johnson once wrote, "What we hope ever to do with ease, we must first learn to do with diligence."[36] Nowhere is this quote more applicable than to Liam and Zohra and their sensory stories. They taught themselves basic perceptual skills that most of us acquired in infancy. Since understanding what we see and hear seems so natural to most of us, we assume that, in infancy, our vision and hearing developed spontaneously. But nothing could be further from the truth. Although, through normal maturation of the eyes, most of us obtained good visual acuity in our first six months, this alone did not allow us to recognize objects. While most of us developed stereovision within the first sixteen weeks, this alone did not allow us to understand the three-dimensional shape of objects and their surroundings. Active exploration and experimentation were required. And as babies, we were highly motivated to experiment and explore. Only by turning our head and eyes toward a sound could we begin to identify what we heard. As we moved our hands back and forth in front of our eyes, we taught ourselves to recognize them from different angles. While swaying our head from side to side, we learned how things look from different points of view. By handling objects, dropping them on the ground, smashing them together, and moving around them, we began to recognize their shapes and properties. As we babbled, we learned what our own utterances sounded like, enabling us to speak our first words. When we began to crawl and walk, we were on the move, exploring the spatial layout. Our parents may have encouraged us in all these activities and provided a stimulating environment, but as babies, we, to a large extent, taught ourselves.[37]

But teaching yourself to see or hear as an adolescent or adult is not so instinctive and automatic.[38] After childhood, we become stuck in our ways. Our daily habits are very useful in guiding us through the day with minimal effort, but these habits also limit new exploration and experimentation. To learn to see and hear, Liam and Zohra had to abandon their habitual ways of doing and being and then, like a young child, teach themselves. They had to analyze what they could and couldn't do and then actively design, practice, and master tasks of increasing difficulty. This process was, as Chorost described it, "an intensely conscious act," requiring the discipline of a professional musician or athlete.[39] But once they learned a skill well, it became more automatic and less effortful, allowing them to relax and take in the broader context.

When I undertook vision therapy, I kept a daily journal because it encouraged me to reflect carefully on my new sensations and experiences. Arlene Romoff, who had great success with her cochlear implants, kept and published two detailed journals.[40] She started out with an implant in only her left ear but, after ten years, received an implant in the right. By this time, she was so experienced in rehabilitating herself and in weathering the emotions that accompany dramatic sensory changes that she knew how to pace her exposure to and design exercises for her bilateral hearing. She called the car her "mobile hearing testing lab" because, while driving, she would listen to the car radio and CD player to practice and evaluate her hearing.

"I'm starting to develop a relationship with my brain, appreciating its power over my hearing," Romoff wrote. With the increased self-awareness that is so important in rehabilitating yourself comes the experience of watching your own brain in action. While listening to children's books on tape, Michael Chorost had the strange experience of initially hearing a growly voice speaking "pseudo-English" that, with practice, morphed into understandable speech. The voice he heard on tape still sounded to him like buzzes and growls, but somehow meaningful language came through. "You neurons," he wrote, "you sort it out in there."[41]

As Liam and Zohra embraced their new senses, they truly became "athletes of perception," training themselves to see or hear like an athlete trains for his sport. No new sensation or experience was left unanalyzed. But why weren't the demands of everyday tasks sufficient to hone Liam's and Zohra's perceptual skills? Why did they have to pay so much attention and give so much thought to how and what they were sensing? In order for their new sensations to be of use, they had to be understood; they had to have meaning and relevance. Only then could Liam and Zohra assimilate them into their overall worldview.

Indeed, scientific experiments suggest that passive exposure to new stimuli may result in very little new learning. Both selective attention to those stimuli and the release of neuromodulators in the brain are required. These neuromodulators, including dopamine and acetylcholine, are liberated when we are alert and exploring our environment, when we are learning about novel stimuli, and when we anticipate a reward for our actions. Combined, selective attention and neuromodulator release facilitate brain plasticity and the setting down of new memories specifically involving those stimuli.[42] Here is a scientific explanation for a phenomenon we've all observed: we learn best what we want to learn. Even as a deaf child, Zohra loved the words in books. Her thrill at first hearing the way a word sounded reinforced the memory of that sound and made it easier for her to recognize the word when she heard it spoken again.

From the time Zohra was an infant, she received auditory training from her beloved Najma. When she was three, Najma wrote in her diary, "[Zohra] shows so much love. Together we work and enjoy every minute of it." As with Anne Sullivan and Helen Keller, lessons became a form of play. For Zohra, learning and therapy were associated with being with Najma, with showing Najma how much she could do. These feelings carried forward into adulthood and gave Zohra the confidence to navigate the world with her own unique sensory makeup.

Since Liam received no formal visual training after first receiving his IOLs, he became, with Cindy's unfailing support and guidance, his own teacher and coach. One day in 2011, Liam and I were discussing his thoughts about "the shape of vision" by exchanging emails using his mother's email address. Even after Liam received his IOLs, he wondered about the limits of a person's peripheral vision. Did the world appear as if it were inside a square or oval frame, like on a TV screen? It was at this point that Cindy jumped into our discussion: "Please pardon my intrusion into your conversation . . . and pardon my emotional perspective intruding on the analytical aspect . . . You guys drive me mad! Yikes! I wish you could feel the wild tumult that runs around inside of me when I read your questions, observations and thoughts! . . . And now, NOW you bring up that question again about the shape of your vision! I'm frantically looking around trying to find a shape for my vision. I don't perceive one. There are no edges." Then Cindy added that she was aware not only of what she can see but of what she knows is there, even if some things are hidden from view. "No need to analyze the perspective of all this, nor to think through what is unseen," Cindy concluded. "It's automatic for me. No wonder sighted people are so squishy and fuzzy, lacking all that analyticalness that comes so easily to you."

That "analyticalness," which both Liam and Zohra share, may have been honed by the way they grew up and the difference in the way they and others perceived the world. Both lived and went to school with people who could see or hear, so they had to constantly imagine and infer what others were sensing. Philosopher R. Latta, in his essay about John Carruth, a blind man who gained sight at age thirty, put this best when he wrote that Carruth had to live in two worlds: the world as he sensed it and the world seen by the people around him.[43] Carruth was an active participant in the latter world. He walked about his hometown with confidence. As a boy he was a leader in games and climbed trees fearlessly. He harvested corn in the fields and delivered goods for a grocer. Through all this, he had to interpret what other people saw. Indeed, Latta suggests that Carruth's powers of imagination and inference were greater than they would have been had he not been blind. And these same powers allowed him, after his operations, to accommodate to and make use of his sight. Latta contrasts Carruth's story with that of Carruth's sister, who was also congenitally blind. The sister was educated in a blind school and had a job that didn't demand sight. When her sight was restored, she hardly made use of it. Robert Hine, blinded for fifteen years, echoed the same sentiments as Latta when he wrote of the blind participating in the world of the sighted: "Whether or not senses like hearing and touch grow sharper, the ability to imagine must intensify, and it is there that the blind can outshine the seeing."[44] The same is true for Liam and Zohra.

Is there a limit to neuronal plasticity and the learning that results? Certainly. At the cellular level, plasticity and learning require the growth and retraction of different neuronal processes, the formation of new synapses and elimination of old ones, adjustments in the strength of existing ones, and changes in the extracellular matrix surrounding those synapses.[45] All this requires time, new cellular structures, and energy. But other factors are also involved. Every time we watch or listen to performers at a circus, symphony, ballet, or professional ball game, we are seeing the results of neuronal plasticity.

While these performers may possess natural gifts, they wouldn't have achieved the elite level without a singular focus and years of intense practice. Neuronal plasticity and learning require active training, and there are limits in terms of time, finances, motivation, and energy to how much we can practice.

How far can Liam and Zohra improve? Both face substantial roadblocks. Because of albinism, Liam's retina and eye-brain wiring did not develop normally, and severe myopia (nearsightedness) deprived him of good acuity for his first fifteen years. Zohra had very little hearing experience as a young child; a cochlear implant does not provide the richness of sound provided by normal ears, and she has an implant in only one ear, which severely hampers her ability to localize sounds. But these limitations do not prevent future growth and improvements as long as Liam's eyes remain healthy and Zohra's implant continues to work.

Liam is intrigued by the case of Michael May, another sight-recovery person.[46] His visual skills were tested three years after sight recovery in 2003 and then again a decade later in 2013. During the intervening ten years, his ability to recognize objects or determine gender or emotion from a person's face had not improved. These results may suggest a lack of plasticity in the adult brain, but that may be only part of the explanation. May wrote, "I have learned what works with vision and what doesn't so I really don't challenge my vision much anymore. This means where motion or color might be clues, I use my vision. Where details might be required, like reading print or recognizing who someone is, I use tactile and auditory techniques."[47]

Like Michael May, all of us settle for a certain level of competence in different tasks. We may develop a skill to the point that it is sufficient but no better than necessary for the task at hand. Children develop skills through their own exploration and play, but once we are grown, we seem stuck in our ways. For many of us, for example, our drawing skills may have peaked in the fourth grade. Frustrated with our childlike pictures, perhaps we chose to develop other skills that were more important to us or more fun. Yet most of us could improve our drawing skills, often to a considerable degree, if we took a drawing class or followed the instructions in a book such as Betty Edwards's *Drawing on the Right Side of the Brain.*[48] The brilliant physicist Richard Feynman learned to draw at age forty-four, and his sketches are surprisingly good.[49] But every new skill requires effort, and we must constantly weigh whether the benefits justify the work involved.

When Liam and Zohra first gained a new sense, they were bombarded with novel sensations—colors and lines, a cacophony of sounds—which were disembodied and meaningless. Both had to integrate these raw stimuli with information from their other senses in order to perceive a world made up not of isolated stimuli but of whole objects and events. Their new sense did not merely sharpen their former worldview but changed it in qualitative ways. Liam could see further and faster. He could see objects in motion and how his view continually changed as he walked. With her cochlear implant, Zohra could hear things she couldn't see, such as sounds produced behind walls or around corners. Both had to retrain the way they took in their surroundings, reorder their perceptual world, and rewire the networks in their brain. In short, they had to learn a new way of being.

Psychologists Eleanor Gibson and Anne Pick write, "As humans, we must be flexible and adaptable to change, but we also strive for economy and efficiency in perceiving the world, in action, and in thinking."[50] Our survival depends upon our ability to adapt and learn. From the time we can walk, we're on the move, encountering new places, people, and things. Even if we stay put, the world around us continually changes. But given the wealth of information out there, we must also be selective about what we attend to. Liam and Zohra understood this best. They adapted to great change, and they accomplished this by being both flexible and efficient. As Liam has pointed out to me several times, his goal is not so much to see better but to use his vision, combined with nonvisual skills, to function better. Given that we use our vision in myriad ways, he recognized early on that he had to prioritize what visual skills he developed. Navigation and sports were most important to him, and this is where he has enjoyed great success. But it was Zohra who put this best when she described an achievement she once considered impossible. After college at Mount Holyoke, she returned to her native Tanzania, where she attended and completed medical school. As a child, she learned English, but as a doctor, she needed to speak Swahili in order to talk with her patients. Thinking back to her experiences, she wrote to me,

> And I think sometimes we underestimate our ability to accomplish things. If someone had told me at Mount Holyoke that in a few years' time I would be conversing with patients in Swahili and taking medical histories from them, I would probably have laughed at that and would have said that it is impossible, since it is very hard for a person who cannot hear to learn a new language. Having grown up and lived in Tanzania for eighteen years, I still hadn't learned Swahili. I think the phrase "necessity is the mother of invention" is true. When I had no choice but to learn Swahili in medical school in order to be able to talk to the patients—that is when I realized how much potential we have—especially when we are pushed out of our comfort zone. The brain learns it somehow.

Acknowledgments

I AM DEEPLY GRATEFUL TO MY BOOK AGENT, LISA ADAMS, OF THE Garamond Agency for her wise advice, steady support, and patient review of many drafts of the book proposal. Eric Henney, my editor at Basic Books, encouraged me to make explicit the essential themes of the book and reminded me always to keep the "big picture" in mind. He helped me turn a collection of stories into a coherent narrative. I am grateful as well to my expert copy editor Jen Kelland and proofreader Carrie Wicks and to Melissa Raymond and Brynn Warriner at Basic for their publishing support. I also thank Chin-Lee Yai for the beautiful cover and Linda Mark for the book's interior design.

Benjamin Backus, Beverly Biderman, Michael Chorost, Paul Harris, Leonard Press, Arlene Romoff, and Lawrence Tychsen read earlier drafts of this book and provided important corrections and observations. Any mistakes that remain are my own. And a special thanks to Michael Chorost for the use of the phrase "athletes of perception" as the title of the concluding chapter. For help with illustrations, I thank James Gehrt, Margaret Nelson, and Andy Barry. Financial support for this book was provided by an Emeriti Grant from Mount Holyoke College.

I thank Lawrence Tychsen, Liam's ophthalmologist, for introducing me to Liam, providing me with medical information on Liam's case, and reviewing an earlier draft of the book. Judy Stockstad helped arrange my first meeting with Liam, at which I also benefitted from discussions with Liam's optometrist, James Hoekel.

As always, I thank my family—my ever-supportive, swashbuckling husband, Dan Barry; my children, Jenny and Andy Barry; their spouses, David German and Katya Kosheleva; my brother, Daniel Feinstein, and sister, Deborah Feinstein; their spouses, Rose Ann Wasserman and Alan Copsey; and Dan's three sisters, Janice Grzesinski, Carol Chiodi, and Kathleen Jackson. I thank my granddaughter, Jessica, for showing me in real life what I had studied in books—how infants learn to perceive the world.

I am grateful to Kate Edgar, Linda Laderach, Andy Lass, John Lemly, Chris Pyle, Bill Quillian, Stan Rachootin, Margaret Robinson, Diana Stein, Al Werner, Ren Weschler, Rosalie Winard, and especially Rachel Fink for their support and friendship.

I owe a great debt to the late Oliver Sacks, who strongly encouraged me to write this book.

Above all, I thank Liam McCoy and Zohra Damji for teaching me so much about resilience and resourcefulness, vision and hearing. They were thoughtful and thorough in the way they told me their stories, generous with their time, and willing to read multiple drafts of the book. I am proud to call them my friends. I thank, too, their families for their openness, insights, and warmth, especially Cindy Lansford and Najma Mussa. It is to Cindy and Najma, for their courage, commitment, and strength, that I dedicate this book.

Figure Credits

Figures not specifically credited below were drawn or photographed by the author.

Figure 1.1 Images cast on the retina: © Margaret Nelson.

Figure 1.2 Optic pathways and optic chiasm: © Margaret Nelson.

Figure 3.2 Liam's sketch of a flower: Drawn by Liam McCoy.

Figure 3.3 Liam's sketch of a cat: Drawn by Liam McCoy.

Figure 3.5 Hering illusion: Wikipedia (https://en.wikipedia.org/wiki/Hering_illusion#/media/File:Hering_illusion.svg), by Fibonacci, licensed under CC BY-SA 3.0 (https://creativecommons.org/licenses/by-sa/3.0).

Figure 3.6 Transparency: From J. Albers, *Interaction of Color* (New Haven, CT: Yale University Press, 2006).

Figure 3.7 A highly simplified schematic of the action and perception pathways: Image by Andrew J. Barry.

Figure 3.8 Rubin's vase: V. Ramachandran and D. Rogers-Ramachandran, "Ambiguities and Perception," *Scientific American Mind* 18 (2007): 18.

Figure 3.9 Wilson's illusion: © 1992 by Bruner/Mazel Inc. From J. R. Block and H. E. Yuker, *Can You Believe Your Eyes?* Reproduced by permission of Taylor and Francis Group, LLC, a division of Informa PLC.

Figure 3.11 Contour integration: Wikipedia (https://commons.wikimedia.org/wiki/File:Contour_Integration_Example_1.jpg), by Mundhenk, licensed under CC BY-SA 3.0 (https://creativecommons.org/licenses/by-sa/3.0).

Figure 3.12 Camouflage: snake in the leaves: From Rockwell Kent, "Copperhead Snake on Dead Leaves," in A. H. Thayer, *Concealing-Coloration in the Animal Kingdom* (New York: Macmillan, 1909). Public domain.

Figure 3.13 Liam's abstract drawing: Drawn by Liam McCoy.

Figure 4.1 *Portrait with Vegetables* by Giuseppe Arcimboldo: Picture Art Collection / Alamy Stock Photo (https://www.alamy.com/stock-photo-giuseppe-arcimboldo-the-vegetable-gardener-91283590.html).

Figure 4.2 Chuck Close, *Self-Portrait*, 2007: © Chuck Close, courtesy Pace Gallery; first viewed at Mount Holyoke College Art Museum, South Hadley, Massachusetts, purchased with the Art Acquisition Endowment Fund.

Figure 5.1 Kanisza triangle: Wikipedia, (https://en.wikipedia.org/wiki/Illusory_contours#/media/File:Kanizsa_triangle.svg), by Fibonacci, licensed under CC BY-SA 3.0 (https://creativecommons.org/licenses/by-sa/3.0).

Figure 5.4 The fragments we see in the left half of the figure take on new meaning when we view them in the right half: Reproduced, with permission, from A. S. Bregman, "Asking the 'What For' Question in Auditory Perception," in M. Kubovy and J. R. Pomerantz, eds., *Perceptual Organization*, (Hillsdale, NJ: Lawrence Erlbaum, 1981).

Figure 5.6 The Quoits vectogram: Photo by James Gehrt.

Figure 8.2 The Ponzo illusion: Wikipedia (https://en.wikipedia.org/wiki/Ponzo_illusion#/media/File:Ponzo_illusion.gif), by Tony Philips, National Aeronautics and Space Administration. Public domain.

Figure 8.4 The corridor illusion: By J. Deregowski in R. L. Gregory and E. H. Gombrich, eds., *Art and Illusion* (London: Duckworth, 1973).

Figure 8.5 Staircases: © James Gehrt.

Figure 8.6 Reflections through a plate-glass window: © James Gehrt.

Figure 8.7 Shading creates a sense of depth: Image by James Gehrt.

Figure 8.8: Shadows affect our interpretation of the balls' location: Image by Andrew J. Barry.

Figure 8.9 "An early bird who caught a very strong worm": Excerpted from *The Ultimate Droodles Compendium* by Roger Price Copyright ©2019 Tallfellow Press, used with permission. All rights reserved.

Figure 9.1 Liam, 2019: Photo by Pixel Caliber Collective.

Figure 10.1 Najma with Zohra on the London Underground: Damji family photos.

Figure 18.1 Zohra, in Toronto, 2019: Damji family photos.

Notes

INTRODUCTION: BLESSING OR CURSE?

1. R. L. Gregory and J. G. Wallace, *Recovery from Early Blindness: A Case Study*, Monograph No. 2 (Cambridge, UK: Experimental Psychology Society, 1963).

2. O. Sacks, "To See and Not See," in *An Anthropologist on Mars: Seven Paradoxical Tales* (New York: Alfred A. Knopf, 1995).

3. B. Biderman, *Wired for Sound: A Journey into Hearing* (Toronto: Journey into Hearing Press, 2016), 26.

4. S. R. Barry, *Fixing My Gaze: A Scientist's Journey into Seeing in Three Dimensions* (New York: Basic Books, 2009).

5. I. Biederman et al., "On the Information Extracted from a Glance at a Scene," *Journal of Experimental Psychology* 103 (1974): 597–600.

6. A. Valvo, *Sight Restoration After Long-Term Blindness: The Problems and Behavior Patterns of Visual Rehabilitation* (New York: American Federation for the Blind, 1971), 39.

7. A. S. Bregman, *Auditory Scene Analysis: The Perceptual Organization of Sound* (Cambridge, MA: MIT Press, 1990).

8. Valvo, *Sight Restoration After Long-Term Blindness*, 9.

9. D. Wright, *Deafness: An Autobiography* (New York: Harper Perennial, 1993), 14.

10. Biderman, *Wired for Sound*; H. Lane, *The Mask of Benevolence: Disabling the Deaf Community* (New York: Knopf, 1992).

11. R. Latta, "Notes on a Case of Successful Operation for Congenital Cataract in an Adult," *British Journal of Psychology* 1 (1904): 135–150.

12. Gregory and Wallace, *Recovery from Early Blindness*.

13. Valvo, *Sight Restoration After Long-Term Blindness*.

14. M. von Senden, *Space and Sight: The Perception of Space and Shape in the Congenitally Blind Before and After Operation* (Glencoe, IL: Free Press, 1960).

15. Gregory and Wallace, *Recovery from Early Blindness*.

16. J. M. Hull, *Touching the Rock: An Experience of Blindness* (New York: Pantheon Books, 1990), 94.

17. Valvo, *Sight Restoration After Long-Term Blindness*, 12.

18. I. Rosenfield, *The Invention of Memory: A New View of the Brain* (New York: Basic Books, 1988).

19. Hull, *Touching the Rock*, 217.

20. E. J. Gibson and A. D. Pick, *An Ecological Approach to Perceptual Learning and Development* (New York: Oxford University Press, 2000); M. E. Arterberry and P. J. Kellman, *Development of Perception in Infancy: The Cradle of Knowledge Revisited* (New York: Oxford University Press, 2016).

21. O. Sacks, *The River of Consciousness* (New York: Knopf, 2017), 183.

22. V. W. Tatler et al., "Yarbus, Eye Movements, and Vision," *i-Perception* (2010): 7–27.

23. J. W. Henderson, "Gaze Control as Prediction," *Trends in Cognitive Sciences* 21 (2017): 15–23.

24. Gibson and Pick, *An Ecological Approach to Perceptual Learning and Development*; Arterberry and Kellman, *Development of Perception in Infancy*; E. J. Gibson, "Perceptual Learning: Differentiation or Enrichment?," in *An Odyssey in Learning and Perception* (Cambridge, MA: MIT Press, 1991); P. J. Kellman and P. Garrigan, "Perceptual Learning and Human Expertise," *Physics of Life Reviews* 6 (2009): 53–84.

25. A. Noë, *Action in Perception* (Cambridge, MA: MIT Press, 2005), 1.

26. Von Senden, *Space and Sight*.

27. Gregory and Wallace, *Recovery from Early Blindness*, 37 (emphasis in the original).

28. Sacks, "To See and Not See," 132–134.

29. Von Senden, *Space and Sight*.

30. D. H. Hubel and T. N. Wiesel, *Brain and Visual Perception: The Story of a 25-Year Collaboration* (Oxford: Oxford University Press, 2005).

31. Biderman, *Wired for Sound*.

32. Valvo, *Sight Restoration After Long-Term Blindness*; von Senden, *Space and Sight*.

33. R. Kurson, *Crashing Through. A True Story of Risk, Adventure, and the Man Who Dared to See* (New York: Random House, 2007).

34. P. Sinha, "Once Blind and Now They See: Surgery in Blind Children from India Allows Them to See for the First Time and Reveals How Vision Works in the Brain," *Scientific American* 309 (2013): 48–55.

CHAPTER 1: HOW FAR IS YOUR VISION?

1. H. Keller, *The Story of My Life: The Restored Edition*, ed. J. Berger (New York: Modern Library, 2004).

2. N. Daw, *Visual Development*, 3rd ed. (New York: Springer, 2014).

3. A. E. Hendrickson, "Primate Foveal Development: A Microcosm of Current Questions in Neurobiology," *Investigative Ophthalmology & Visual Science* 35 (1994): 3129–3133.

4. G. Jeffery, "The Retinal Pigment Epithelium as a Developmental Regulator of the Neural Retina," *Eye* 12 (1998): 499–503.

5. Hendrickson, "Primate Foveal Development."

6. See also R. Latta, "Notes on a Case of Successful Operation for Congenital Cataract in an Adult," *British Journal of Psychology* 1 (1904): 135–150.

7. Daw, *Visual Development*.

8. Daw, *Visual Development*; K. Apkarian, "Chiasmal Crossing Defects in Disorders of Binocular Vision," *Eye* 10 (1996): 222–232.

9. E. A. H. von dem Hagen et al., "Pigmentation Predicts the Shift in the Line of Decussation in Humans with Albinism," *European Journal of Neuroscience* 25 (2007): 503–511.

10. P. Apkarian and D. Reits, "Global Stereopsis in Human Albinos," *Vision Research* 29 (1989): 1359–1370; A. B. Cobo-Lewis et al., "Poor Stereopsis Can Support Size Constancy in Albinism," *Investigative Ophthalmology & Visual Science* 38 (1997): 2800–2809; K. A. Lee, R. A. King, and C. G. Summers, "Stereopsis in Patients with Albinism: Clinical Correlates," *Journal of AAPOS* 5 (2001): 98–104.

11. K. T. Mullen and F. A. A. Kingdom, "Differential Distributions of Red-Green and Blue-Yellow Cone Opponency Across the Visual Field," *Visual Neuroscience* 19 (2002): 109–118.

CHAPTER 2: DR. RIDLEY'S BRAINCHILD

1. D. J. Apple, *Sir Harold Ridley and His Fight for Sight: He Changed the World So That We May Better See It* (Thorofare, NJ: SLACK, Inc., 2006); D. J. Apple,

"Nicholas Harold Lloyd Ridley, 10 July 1906–25 May 2001: Elected FRS 1986," *Biographical Memoirs of Fellows of the Royal Society* 53 (2007): 285–307.

2. P. U. Fechner, G. L. van der Heijde, and J. G. Worst, "The Correction of Myopia by Lens Implantation into Phakic Eyes," *American Journal of Ophthalmology* 107 (1989): 659–663; J. G. Worst, G. van der Veen, and L. I. Los, "Refractive Surgery for High Myopia: The Worst-Fechner Biconcave Iris Claw Lens," *Documental Ophthalmologica* 75 (1990): 335–341.

3. L. Tychsen, "Refractive Surgery for Special Needs Children," *Archives of Ophthalmology* 127 (2009): 810–813.

4. L. Tychsen et al., "Phakic Intraocular Lens Correction of High Ametropia in Children with Neurobehavioral Disorders," *Journal of AAPOS* 12 (2008): 282–289.

CHAPTER 3: A WINDOW ON THE BRAIN

1. S. Thorpe, D. Fize, and C. Marlot, "Speed of Processing in the Human Visual System," *Nature* 381 (1996): 520–522.

2. A. Valvo, *Sight Restoration After Long-Term Blindness: The Problems and Behavior Patterns of Visual Rehabilitation* (New York: American Federation for the Blind, 1971); M. Von Senden, *Space and Sight: The Perception of Space and Shape in the Congenitally Blind Before and After Operation* (Glencoe, IL: Free Press, 1960).

3. R. Kurson, *Crashing Through: A True Story of Risk, Adventure, and the Man Who Dared to See* (New York: Random House, 2007).

4. S. Hocken, *Emma and I: The Beautiful Labrador Who Saved My Life* (London: Ebury Press, 2011).

5. R. L. Gregory and J. G. Wallace, *Recovery from Early Blindness: A Case Study*, Monograph No. 2 (Cambridge, UK: Experimental Psychology Society, 1963); O. Sacks, "To See and Not See," in *An Anthropologist on Mars: Seven Paradoxical Tales* (New York: Alfred A. Knopf, 1995).

6. R. V. Hine, *Second Sight* (Berkeley: University of California Press, 1993).

7. S. Bitgood, "Museum Fatigue: A Critical Review," *Visitor Studies* 12 (2009): 93–111.

8. D. O. Hebb, *The Organization of Behavior: A Neuropsychological Theory* (Mahwah, NJ: Lawrence Erlbaum Associates, Publishers, 2002).

9. Valvo, *Sight Restoration After Long-Term Blindness*, 39.

10. Y. Ostrovsky et al., "Visual Parsing After Recovery from Blindness," *Psychological Science* 20 (2009): 1484–1491; P. Sinha, "Once Blind and Now They See," *Scientific American* 309 (2013): 48–55; R. Sikl et al., "Vision After 53 Years of Blindness," *i-Perception* 4 (2013): 498–507.

11. Gregory and Wallace, *Recovery from Early Blindness*; Sikl et al., "Vision After 53 Years of Blindness."

12. Valvo, *Sight Restoration After Long-Term Blindness*; T. Gandhi et al., "Immediate Susceptibility to Visual Illusions After Sight Onset," *Current Biology* 25: (2015): R345–R361.

13. Gregory and Wallace, *Recovery from Early Blindness*.

14. P. C. Quinn, P. D. Eimas, and M. J. Tarr, "Perceptual Categorization of Cat and Dog Silhouettes by 3- to 4-Month-Old Infants," *Journal of Experimental Child Psychology* 79 (2001): 78–94.

15. J. Albers, *Interaction of Color*, rev. ed. (New Haven, CT: Yale University Press, 1975).

16. N. Daw, *How Vision Works: The Physiological Mechanisms Behind What We See* (New York: Oxford University Press, 2012).

17. D. H. Hubel and T. N. Wiesel, *Brain and Visual Perception: The Story of a 25-Year Collaboration* (Oxford: Oxford University Press, 2005).

18. C. D. Gilbert and W. Li, "Top-Down Influences on Visual Processing," *Nature Review Neuroscience* 14 (2013): 350–363; W. Li, V. Piëch, and C. D. Gilbert, "Learning to Link Visual Contours," *Neuron* 57 (2008): 442–451.

19. A. R. Luria, *The Working Brain: An Introduction to Neuropsychology* (New York: Basic Books, 1973).

20. Luria, *The Working Brain*; M. J. Farah, *Visual Agnosia*, 2nd ed. (Cambridge, MA: MIT Press, 2004).

21. E. Goldberg, *Creativity: The Human Brain in the Age of Innovation* (New York: Oxford University Press, 2018).

22. O. Sacks, *The Man Who Mistook His Wife for a Hat and Other Clinical Tales* (New York: Summit Books, 1985).

23. Daw, *How Vision Works*.

24. M. A. Goodale and A. D. Milner, "Separate Visual Pathways for Perception and Action," *Trends in Neuroscience* 15 (1992): 20–25.

25. Daw, *How Vision Works*.

26. S. Hochstein and M. Ahissar, "View from the Top: Hierarchies and Reverse Hierarchies in the Visual System," *Neuron* 36 (2002): 791–804.

27. M. E. Arterberry and P. J. Kellman, *Development of Perception in Infancy: The Cradle of Knowledge Revisited* (New York: Oxford University Press, 2016).

28. Hochstein and Ahissar, "View from the Top."

29. S. Hochstein, "The Gist of Anne Triesman's Revolution," *Attention, Perception & Psychophysics*, September 16, 2019, https://doi.org/10.3758/s13414-019-01797-2.

30. J. L. Pind, *Edgar Rubin and Psychology in Denmark: Figure and Ground* (Cham, Switzerland: Springer International Publishing, 2014).

31. V. A. F. Lamme, "The Neurophysiology of Figure-Ground Segregation in Primary Visual Cortex," *Journal of Neuroscience* 15 (1995): 1605–1615.

32. M. Wertheimer, "Laws of Organization in Perceptual Forms," in *A Source Book of Gestalt Psychology*, ed. W. Ellis (London: Routledge & Kegan Paul, 1938), 71–88. First published as "Untersuchungen zur Lehre von der Gestalt II," *Psycologische Forschung* 4 (1923): 301–350.

33. C. F. Altmann, H. H. Bülthoff, and Z. Kourtzi, "Perceptual Organization of Local Elements into Global Shapes in the Human Visual Cortex," *Current Biology* 13 (2003): 342–349; R. E. Crist, W. Li, and C. D. Gilbert, "Learning to See: Experience and Attention in Primary Visual Cortex," *Nature Neuroscience* 4 (2001): 515–525; F. T. Qui, T. Sugihara, and R. von der Heydt, "Figure-Ground Mechanisms Provide Structure for Selective Attention," *Nature Neuroscience* 10 (2007): 1492–1499; F. T. Qui and R. von der Heydt, "Figure and Ground in the Visual Cortex: V2 Combines Stereoscopic Cues with Gestalt Rules," *Neuron* 47 (2005): 155–166.

34. Goldberg, *Creativity*.

35. A. T. Morgan, L. S. Petro, and L. Muckli, "Scene Representations Conveyed by Cortical Feedback to Early Visual Cortex Can Be Described by Line Drawings," *Journal of Neuroscience* 39 (2019): 9410–9423.

36. E. J. Gibson, "Perceptual Learning: Differentiation or Enrichment?," in *An Odyssey in Learning and Perception* (Cambridge, MA: MIT Press, 1991); E. J. Gibson and A. D. Pick, *An Ecological Approach to Perceptual Learning and Development* (New York: Oxford University Press, 2000); P. J. Kellman and P. Garrigan, "Perceptual Learning and Human Expertise," *Physics of Life Reviews* 6 (2009): 53–84.

37. M. Sigman et al., "Top-Down Reorganization of Activity in the Visual Pathway After Learning a Shape Identification Task," *Neuron* 46 (2005): 823–845.

CHAPTER 4: FACES

1. A. W. Young, D. Hellawell, and D. C. Hay, "Configurational Information in Face Perception," *Perception* 16 (1987): 747–759.

2. D. G. Pelli, "Close Encounters—an Artist Shows That Size Affects Shape," *Science* 285 (1999): 844–846; P. Cavanagh and J. M. Kennedy, "Close Encounters: Details Veto Depth from Shadows," *Science* 287 (2000): 2423–2425.

3. "Conversation: Chuck Close, Christopher Finch," *NewsHour*, PBS, July 2, 2010, http://www.pbs.org/newshour/art/conversation-chuck-close-christopher-finch.

4. S. Hocken, *Emma and I: The Beautiful Labrador Who Saved My Life* (London: Ebury Press, 2011), 270.

5. R. L. Gregory and J. G. Wallace, *Recovery from Early Blindness: A Case Study*, Monograph No. 2 (Cambridge, UK: Experimental Psychology Society, 1963); R. Kurson, *Crashing Through: A True Story of Risk, Adventure, and the Man Who Dared to See* (New York: Random House, 2007); O. Sacks, "To See and Not See," in *An Anthropologist on Mars: Seven Paradoxical Tales* (New York: Alfred A. Knopf, 1995); A. Valvo, *Sight Restoration After Long-Term Blindness: The Problems and Behavior Patterns of Visual Rehabilitation* (New York: American Federation for the Blind, 1971); M. Von Senden, *Space and Sight: The Perception of Space and Shape in the Congenitally Blind Before and After Operation* (Glencoe, IL: Free Press, 1960).

6. S. Geldart et al., "The Effect of Early Visual Deprivation on the Development of Face Processing," *Developmental Science* 5 (2002): 490–501; R. A. Robbins et al., "Deficits in Sensitivity to Spacing After Early Visual Deprivation in Humans: A Comparison of Human Faces, Monkey Faces, and Houses," *Developmental Psychobiology* 52 (2010): 775–781.

7. M. E. Arterberry and P. J. Kellman, *Development of Perception in Infancy: The Cradle of Knowledge Revisited* (New York: Oxford University Press, 2016); C. C. Goren, M. Sarty, and P. Y. K. Wu, "Visual Following and Pattern Discrimination of Face-Like Stimuli by Newborn Infants," *Pediatrics* 56 (1975): 544–549; A. Slater, "The Competent Infant: Innate Organization and Early Learning in Infant Visual Perception," in *Perceptual Development: Visual, Auditory, and Speech Perception in Infancy*, ed. A. Slater (East Sussex, UK: Psychology Press Ltd., Publishers, 1998).

8. Arterberry and Kellman, *Development of Perception in Infancy*; I. W. R. Bushnell, F. Sai, and J. T. Mullin, "Neonatal Recognition of the Mother's Face," *British Journal of Developmental Psychology* 7 (1989): 3–15.

9. N. Kanwisher and G. Yovel, "The Fusiform Face Area: A Cortical Region Specialized for the Perception of Faces," *Philosophical Transactions of the Royal Society B* 1476 (2006): 2109–2128.

10. M. Bilalic et al., "Many Faces of Expertise: Fusiform Face Area in Chess Experts and Novices," *Journal of Neuroscience* 31 (2011): 10206–10214.

11. Bilalic et al., "Many Faces of Expertise."

12. C. Turati et al., "Newborns' Face Recognition: Role of Inner and Outer Facial Features," *Child Development* 77 (2006): 297–311.

13. R. Adolphs et al., "A Mechanism for Impaired Fear Recognition After Amygdala Damage," *Nature* 433 (2005): 68–72.

14. Hocken, *Emma and I*.

CHAPTER 5: FINDING THINGS

1. G. Kanisza, "Subjective Contours," *Scientific American* 234 (1976): 48–52.

2. A. L. Bregman, "Asking the 'What For' Question in Auditory Perception," in *Perceptual Organization*, ed. M. Kubovy and J. R. Pomerantz (Hillsdale, NJ: Lawrence Earlbaum, 1981); K. Nakayama and S. Shimojo, "Toward a Neural Understanding of Visual Surface Representation," *The Brain, Cold Spring Harbor Symposium in Quantitative Biology* 55 (1990): 911–924.

3. S. R. Barry, *Fixing My Gaze: A Scientist's Journey into Seeing in Three Dimensions* (New York: Basic Books, 2009).

4. E. E. Birch, S. Shimojo, and R. Held, "Preferential-Looking Assessment of Fusion and Stereopsis in Infants Aged 1–6 Months," *Investigative Ophthalmology & Visual Science* 26 (1985): 366–370; R. Fox et al., "Stereopsis in Human Infants," *Science* 207 (1980): 323–324; B. Petrig et al., "Development of Stereopsis and Cortical Binocularity in Human Infants: Electrophysiological Evidence," *Science* 213 (1981): 1402–1405; F. Thorn et al. "The Development of Eye Alignment, Convergence, and Sensory Binocularity in Young Infants," *Investigative Ophthalmology & Visual Science* 35 (1994): 544–553.

5. M. E. Arteberry and P. J. Kellman, *Development of Perception in Infancy: The Cradle of Knowledge Revisited* (New York: Oxford University Press, 2016); M. Arterberry, A. Yonas, and A. S. Bensen, "Self-Produced Locomotion and the Development of Responsiveness to Linear Perspective and Texture Gradients," *Developmental Psychology* 25 (1989): 976–982; M. Kavsek, A. Yonas, and C. E. Granrud, "Infants' Sensitivity to Pictorial Depth Cues: A Review and Meta-analysis of Looking Studies," *Infant Behavior and Development* 35 (2012): 109–128; A. Tsuruhara et al., "The Development of the Ability of Infants to Utilize Static Cues to Create and Access Representations of Object Shape," *Journal of Vision* 10 (2010), doi:10.1167/10.12.2; A. Yonas and C. E. Granrud, "Infants' Perception of Depth from Cast Shadows," *Perception & Psychophysics* 68 (2006): 154–160.

6. H. Wallach and A. O'Leary, "Slope of Regard as a Distance Cue," *Perception & Psychophysics* 31 (1982): 145–148; A. M. Norcia et al., "Experience-Expectant Development of Contour Integration Mechanisms in Human Visual Cortex," *Journal of Vision* 5 (2005): 116–130.

7. M. von Senden, *Space and Sight: The Perception of Space and Shape in the Congenitally Blind Before and After Operation* (Glencoe, IL: Free Press, 1960).

8. B. Tversky, *Mind in Motion: How Action Shapes Thought* (New York: Basic Books, 2019).

9. S. Hochstein and M. Ahissar, "View from the Top: Hierarchies and Reverse Hierarchies in the Visual System," *Neuron* 36 (2002): 791–804.

10. Von Senden, *Space and Sight*; R. L. Gregory and J. G. Wallace, *Recovery from Early Blindness: A Case Study*, Monograph No. 2 (Cambridge, UK: Experimental Psychology Society, 1963); E. Huber et al., "A Lack of Experience-Dependent Plasticity After More Than a Decade of Recovered Sight," *Psychological Science* 26 (2015): 393–401; O. Sacks, "To See and Not See," in *An Anthropologist on Mars: Seven Paradoxical Tales* (New York: Alfred A. Knopf, 1995); A. Valvo, *Sight Restoration After Long-Term Blindness: The Problems and Behavior Patterns of Visual Rehabilitation* (New York: American Federation for the Blind, 1971).

CHAPTER 6: VISION'S GREATEST TEACHER

1. O. Sacks, "To See and Not See," in *An Anthropologist on Mars: Seven Paradoxical Tales* (New York: Alfred A. Knopf, 1995).

2. R. Kurson, *Crashing Through: A True Story of Risk, Adventure, and the Man Who Dared to See* (New York: Random House, 2007).

3. M. E. Arterberry and P. J. Kellman, *Development of Perception in Infancy: The Cradle of Knowledge Revisited* (New York: Oxford University Press, 2016).

4. Arterberry and Kellman, *Development of Perception in Infancy*; K. J. Kellman and E. S. Spelke, "Perception of Partly Occluded Objects in Infancy," *Cognitive Psychology* 15 (1983): 483–524.

5. M. Wertheimer, "Laws of Organization in Perceptual Forms," in *A Source Book of Gestalt Psychology*, ed. W. Ellis (London: Routledge & Kegan Paul, 1938), 71–88. First published as "Untersuchungen zur Lehre von der Gestalt II," *Psycologische Forschung* 4 (1923): 301–350.

6. B. Tversky, *Mind in Motion: How Action Shapes Thought* (New York: Basic Books, 2019).

7. J. J. Gibson, *The Ecological Approach to Visual Perception* (Hillsdale, NJ: Lawrence Erlbaum Associates, Publishers, 1986).

8. A. Michotte, *The Perception of Causality* (New York: Basic Books, 1963).

9. Arterberry and Kellman, *Development of Perception in Infancy*.

10. Gibson, *The Ecological Approach to Visual Perception*; R. Arnheim, *Visual Thinking* (Berkeley: University of California Press, 1969); H. Wallach and D. N. O'Connell, "The Kinetic Depth Effect," *Journal of Experimental Psychology* 45 (1953): 205–217; E. J. Ward, L. Isik, and M. M. Chun, "General Transformations

of Object Representations in Human Visual Cortex," *Journal of Neuroscience* 38 (2018): 8526–8537. Philosopher Alva Noë maintains that our perception, our understanding of what we sense, is not given to us all at once but comes about through movement and active exploration, even through actions as subtle as our eye movements. A. Noë, *Action in Perception* (Cambridge, MA: MIT Press, 2004).

11. Kurson, *Crashing Through*; I. Fine et al., "Long-Term Deprivation Affects Visual Perception and Cortex," *Nature Neuroscience* 6 (2003): 915–916; Y. Ostrovsky et al., "Visual Parsing After Recovery from Blindness," *Psychological Science* 20 (2009): 1484–1491.

12. P. J. Kellman, "Perception of Three-Dimensional Form by Human Infants," *Perception & Psychophysics* 36 (1985): 353–358.

13. S. Grossberg, "The Resonant Brain: How Attentive Conscious Seeing Regulates Action Sequences That Interact with Attentive Cognitive Learning, Recognition, and Prediction," *Attention, Perception & Psychophysics* 81 (2019): 2237–2264.

14. C. Von Hofsten, "Predictive Reaching for Moving Objects by Human Infants," *Journal of Experimental Child Psychology* 30 (1980): 369–382.

15. M. Dadarlat and M. P. Stryker, "Locomotion Enhances Neural Encoding of Visual Stimuli in Mouse V1," *Journal of Neuroscience* 37 (2017): 3764–3775.

16. T. Bullock et al., "Acute Exercise Modulates Feature-Selective Responses in Human Cortex," *Journal of Cognitive Neuroscience* 29 (2017): 605–618.

17. M. Kaneko, Y. Fu, and M. P. Stryker, "Locomotion Induces Stimulus-Specific Response Enhancement in Adult Visual Cortex," *Journal of Neuroscience* 37 (2017): 3532–3543; M. Kaneko and M. P. Stryker, "Sensory Experience During Locomotion Promotes Recovery of Function in Adult Visual Cortex," *eLife* (2014): 3e02798; C. Lunghi and A. Sale, "A Cycling Lane for Brain Rewiring," *Current Biology* 25 (2015): R1122–R1123.

18. Ostrovsky et al., "Visual Parsing After Recovery from Blindness"; P. Sinha, "Once Blind and Now They See: Surgery in Blind Children from India Allows Them to See for the First Time and Reveals How Vision Works in the Brain," *Scientific American* 309 (2013): 48–55.

CHAPTER 7: GOING WITH THE FLOW

1. E. Nawrot, S. I. Mayo, and M. Nawrot, "The Development of Depth Perception from Motion Parallax in Infancy," *Attention, Perception & Psychophysics* 71 (2009): 194–199; E. Nawrot and M. Nawrot, "The Role of Eye

Movements in Depth from Motion Parallax During Infancy," *Journal of Vision* 13 (2013): 1–13.

2. J. J. Gibson, *The Ecological Approach to Visual Perception* (Hillsdale, NJ: Lawrence Erlbaum Associates, Publishers, 1986).

3. S. L. Strong et al., "Differential Processing of the Direction and Focus of Expansion of Optic Flow Stimuli in Areas MST and V3A of the Human Visual Cortex," *Journal of Neurophysiology* 117 (2017): 2209–2217; R. H. Wurtz and C. J. Duffy, "Neural Correlates of Optic Flow Stimulation," *Annals of the New York Academy of Sciences* 656 (1992): 205–219.

4. F. A. Miles, "The Neural Processing of 3-D Visual Information: Evidence from Eye Movements," *European Journal of Neuroscience* 10 (1998): 811–822.

5. S. Hocken, *Emma and I: The Beautiful Labrador Who Saved My Life* (London: Ebury Press, 2011).

6. S. Holcomb and S. Eubanks, *But Now I See: My Journey from Blindness to Olympic Gold* (Dallas, TX: Benbella Books, Inc., 2013).

CHAPTER 8: FINDING HIS WAY

1. S. Hocken, *Emma and I: The Beautiful Labrador Who Saved My Life* (London: Ebury Press, 2011), 149.

2. J. McPhee, *A Sense of Where You Are: A Profile of William Warren Bradley* (New York: Farrar, Straus and Giroux, 1978), 22.

3. E. C. Tolman, "Cognitive Maps in Rats and Men," *Psychological Review* 55 (1948): 180–208.

4. K. Lorenz, *Here Am I—Where Are You?: The Behavior of the Greylag Goose* (New York: Harcourt Brace Jovanovich, 1988), 18–20.

5. For evidence that spatial skills are not dependent on vision, see R. L. Klatsky et al., "Performance of Blind and Sighted Persons on Spatial Tasks," *Journal of Visual Impairment & Blindness* 89 (1995): 70–82.

6. R. M. Grieve and K. J. Jeffery, "The Representation of Space in the Brain," *Behavioural Processes* 135 (2017): 113–131; C. G. Kentros et al., "Increased Attention to Spatial Context Increases Both Place Field Stability and Spatial Memory," *Neuron* 42 (2004): 283–295; J. O'Keefe, and L. Nadel, *The Hippocampus as a Cognitive Map* (Oxford: Oxford University Press, 1978).

7. M. E. Arteberry and P. J. Kellman, *Development of Perception in Infancy: The Cradle of Knowledge Revisited* (New York: Oxford University Press, 2016); M. Arterberry, A. Yonas, and A. S. Bensen, "Self-Produced Locomotion and the Development of Responsiveness to Linear Perspective and Texture

Gradients," *Developmental Psychology* 25 (1989): 976–982; M. Kavsek, A. Yonas, and C. E. Granrud, "Infants' Sensitivity to Pictorial Depth Cues: A Review and Meta-analysis of Looking Studies," *Infant Behavior and Development* 35 (2012): 109–128; A. Tsuruhara et al., "The Development of the Ability of Infants to Utilize Static Cues to Create and Access Representations of Object Shape," *Journal of Vision* 10 (2010), doi:10.1167/10.12.2; A. Yonas and C. E. Granrud, "Infants' Perception of Depth from Cast Shadows," *Perception & Psychophysics* 68 (2006): 154–160.

8. J. J. Gibson, *The Ecological Approach to Visual Perception* (Hillsdale, NJ: Lawrence Erlbaum Associates Publishers, 1986).

9. Gibson, *The Ecological Approach to Visual Perception.*

10. Curiously, children in Project Prakash were fooled by the Ponzo illusion when they were tested within forty-eight hours after cataract operations that enabled them to see. They saw the two gray lines as different sizes: T. Gandhi et al., "Immediate Susceptibility to Visual Illusions After Sight Onset," *Current Biology* 25 (2015): R345–R361.

11. O. Sacks, "To See and Not See," in *An Anthropologist on Mars: Seven Paradoxical Tales* (New York: Alfred A. Knopf, 1995), 120–121.

12. V. S. Ramachandran, "Perceiving Shape from Shading," *Scientific American* 259 (1988): 76–83.

13. M. Von Senden, *Space and Sight: The Perception of Space and Shape in the Congenitally Blind Before and After Operation* (Glencoe, IL: Free Press, 1960).

14. Arteberry and Kellman, *Development of Perception in Infancy*; Arterberry, Yonas, and Bensen, "Self-Produced Locomotion"; Kavsek, Yonas, and Granrud, "Infants' Sensitivity to Pictorial Depth Cues"; Tsuruhara et al., "The Development of the Ability of Infants to Utilize Static Cues"; Yonas and Granrud, "Infants' Perception of Depth from Cast Shadows."

CHAPTER 10: EVERYTHING HAS A NAME

1. This quote is attributed to Helen Keller, though its actual source has not been found. Keller, however, did express the same idea at other times. For example, in *Helen Keller in Scotland: A Personal Record Written by Herself*, ed. James Kerr Love (London: Methuen & Co., 1933), she wrote, "The problems of deafness are deeper and more complex, if not more important, than those of blindness. Deafness is a much worse misfortune. For it means the loss of the most vital stimulus—the sound of the voice that brings language, sets

thoughts astir and keeps us in the intellectual company of man." See "FAQ: Deaf People in History: Quotes by Helen Keller," Gallaudet University, http://libguides.gallaudet.edu/c.php?g=773975&p=5552566.

2. D. Wright, *Deafness: An Autobiography* (New York: Harper Perennial, 1993).

3. H. Keller, *The Story of My Life: The Restored Edition*, ed. J. Berger (New York: Modern Library, 2004).

4. S. Schaller, *A Man Without Words* (Berkeley: University of California Press, 1991).

5. M. Chorost, *Rebuilt: How Becoming Part Computer Made Me More Human* (New York: Houghton Mifflin, 2005), 31.

6. L. Vygotsky, *Thought and Language*, ed. Alex Kozulin (Cambridge, MA: MIT Press, 1986).

7. J. Bruner, *Child's Talk* (New York: W. W. Norton & Co., Inc., 1983).

8. O. Sacks, *Seeing Voices: A Journey into the World of the Deaf* (Berkeley: University of California Press, 1989).

9. Keller, *The Story of My Life*, 262.

10. J. Rosner, *If a Tree Falls: A Family's Quest to Hear and Be Heard* (New York: Feminist Press, 2010), 65.

11. Keller, *The Story of My Life*.

12. Keller, *The Story of My Life*, 49.

13. S. Hochstein and M. Ahissar, "View from the Top: Hierarchies and Reverse Hierarchies in the Visual System," *Neuron* 36 (2002): 791–804.

CHAPTER 11: PERSISTENCE PAYS OFF

1. B. S. Wilson and M. F. Dorman, "Cochlear Implants: A Remarkable Past and a Brilliant Future," *Hearing Research* 242 (2008): 3–21; A. A. Eshraghi et al., "The Cochlear Implant: Historical Aspects and Future Prospects," *Anatomical Record* 295 (2012): 1967–1980.

2. Wilson and Dorman, "Cochlear Implants"; Eshraghi et al., "The Cochlear Implant"; W. F. House, *The Struggles of a Medical Innovator: Cochlear Implants and Other Ear Surgeries* (William F. House, DDS, MD, 2011).

3. House, *The Struggles of a Medical Innovator*.

4. Wilson and Dorman, "Cochlear Implants"; Eshraghi et al., "The Cochlear Implant."

5. House, *The Struggles of a Medical Innovator*; G. Clark, *Sounds from Silence: Graeme Clark and the Bionic Ear Story* (Crows Nest NSW, Australia: Allen and Unwin, 2000).

6. Wilson and Dorman, "Cochlear Implants."

7. Wilson and Dorman, "Cochlear Implants"; Eshraghi et al., "The Cochlear Implant"; R. C. Bilger and F. O. Black, "Auditory Prostheses in Perspective," *Annals of Otology, Rhinology, and Laryngology* 86, no. 3 (suppl) (May 1977): 3–10, doi:10.1177/00034894770860S301.

8. House, *The Struggles of a Medical Innovator*; B. Biderman, *Wired for Sound: A Journey into Hearing*, rev. ed. (Toronto: Journey into Hearing Press, 2016); H. Lane, *The Mask of Benevolence: Disabling the Deaf Community* (New York: Knopf, 1992).

9. O. Sacks, *Seeing Voices: A Journey into the World of the Deaf* (Berkeley: University of California Press, 1989); D. Wright, *Deafness: An Autobiography* (New York: Harper Perennial, 1993).

10. Biderman, *Wired for Sound*; Lane, *The Mask of Benevolence.*

11. Clark, *Sounds from Silence.*

12. Michael Chorost provides an excellent description of the workings of the cochlear implant in *Rebuilt: How Becoming Part Computer Made Me More Human* (New York: Houghton Mifflin, 2005).

CHAPTER 12: AN UNCANNY FEELING

1. O. Sacks, "To See and Not See," in *An Anthropologist on Mars: Seven Paradoxical Tales* (New York: Alfred A. Knopf, 1995).

2. M. E. Arterberry and P. J. Kellman, *Development of Perception in Infancy: The Cradle of Knowledge Revisited* (New York: Oxford University Press, 2016).

3. D. Maurer, L. C. Gibson, and F. Spector, "Infant Synaesthesia: New Insights into the Development of Multisensory Perception," in *Multisensory Development*, ed. A. J. Bremner, D. J. Lewkowicz, and C. Spence (Oxford: Oxford University Press, 2012).

4. A. Damasio, *Descartes' Error: Emotion, Reason, and the Human Brain* (New York: Penguin Books, 1994), (emphasis in the original).

5. J. J. Gibson, *The Ecological Approach to Visual Perception* (Hillsdale, NJ: Lawrence Erlbaum Associates, Inc., 1986), 116.

CHAPTER 13: SQUEAKS, BANGS, AND LAUGHTER

1. J. M. Hull, *Touching the Rock: An Experience of Blindness* (New York: Pantheon Books, 1990), 82.

2. J. Rosner, *If a Tree Falls: A Family's Quest to Hear and Be Heard* (New York: Feminist Press, 2010).

3. J. Schnupp, I. Nelken, and A. J. King, *Auditory Neuroscience: Making Sense of Sound* (Cambridge, MA: MIT Press, 2012).

4. G. Chechik and I. Nelken, "Auditory Abstraction from Spectro-temporal Features to Coding Auditory Entities," *Proceedings of the National Academy of Sciences* 109 (2012): 18968–18973; L. J. Press, *Parallels Between Auditory and Visual Processing* (Santa Ana, CA: Optometric Extension Program Foundation Inc., 2012).

5. A. R. Luria, *The Working Brain: An Introduction to Neuropsychology* (New York: Basic Books, 1973).

6. A. Bregman, *Auditory Scene Analysis: The Perceptual Organization of Sound* (Cambridge, MA: MIT Press, 1990).

7. Schnupp, Nelken, and King, *Auditory Neuroscience*; Bregman, *Auditory Scene Analysis*.

8. M. Ahissar et al., "Reverse Hierarchies and Sensory Learning," *Philosophical Transactions of the Royal Society B* 364 (2009): 285–299; M. Nahun, I. Nelken, and M. Ahissar, "Stimulus Uncertainty and Perceptual Learning: Similar Principles Govern Auditory and Visual Learning," *Vision Research* 50 (2010): 391–401.

9. The novel sounds of the wind blowing, potato chips crunching, or something falling to the floor were also noted by others when they first received a cochlear implant. B. Biderman, *Wired for Sound: A Journey into Hearing*, rev. ed. (Toronto: Journey into Hearing Press, 2016); A. Romoff, *Hear Again: Back to Life with a Cochlear Implant* (New York: League for the Hard of Hearing, 1999).

10. A. Storr, *Music and the Mind* (New York: Free Press, 1992).

11. M. W. Kraus, "Voice-Only Communication Enhances Empathic Accuracy," *American Psychologist* 72 (2017): 644–654; J. Zaki, N. Bolger, and K. Ochsner, "Unpacking the Informational Bases of Empathic Accuracy," *Emotion* 9 (2009): 478–487.

12. M. D. Pell et al., "Preferential Decoding of Emotion from Human Non-linguistic Vocalizations Versus Speech Prosody," *Biological Psychology* 111 (2015): 14–25.

13. S. Horowitz, *The Universal Sense: How Hearing Shapes the Mind* (New York: Bloomsbury, 2013).

14. S. Manninen et al., "Social Laughter Triggers Endogenous Opioid Release in Humans," *Journal of Neuroscience* 37 (2017): 6125–6131.

15. G. Concina et al., "The Auditory Cortex and the Emotional Valence of Sounds," *Neuroscience and Biobehavioral Reviews* 98 (2019): 256–264.

CHAPTER 14: TALKING TO OTHERS

1. J. Schnupp, I. Nelken, and A. J. King, *Auditory Neuroscience: Making Sense of Sound* (Cambridge, MA: MIT Press, 2012).

2. A. Romoff, *Hear Again: Back to Life with a Cochlear Implant* (New York: League for the Hard of Hearing, 1999).

3. Schnupp, Nelken, and King, *Auditory Neuroscience.*

4. Romoff, *Hear Again.*

5. J. J. Gibson, *The Ecological Approach to Visual Perception* (Hillsdale, NJ: Lawrence Erlbaum Associates, Publishers, 1986).

6. D. Wright, *Deafness: An Autobiography* (New York: Harper Perennial, 1993).

7. M. Chorost, *Rebuilt: How Becoming Part Computer Made Me More Human* (New York: Houghton Mifflin, 2005), 90–91.

8. Wright, *Deafness.*

9. W. T. Gallwey, *The Inner Game of Tennis: The Classic Guide to the Mental State of Peak Performance* (New York: Random House, 1997).

10. T. J. Rogers, B. L. Alderman, and D. M. Landers, "Effects of Life-Event Stress and Hardiness on Peripheral Vision in a Real-Life Stress Situation," *Behavioral Medicine* 29 (2003): 21–26.

11. Romoff, *Hear Again.*

CHAPTER 15: TALKING TO HERSELF

1. L. Vygotsky, *Thought and Language* (Cambridge, MA: MIT Press, 1986).

2. D. Wright, *Deafness: An Autobiography* (New York: Harper Perennial, 1993).

3. R. Arnheim, *Visual Thinking* (Berkeley: University of California Press, 1969).

4. B. Tversky, *Mind in Motion: How Action Shapes Thought* (New York: Basic Books, 2019).

5. M. Schafer and D. Schiller, "In Search of the Brain's Social Road Maps," *Scientific American* 322 (2020): 30–35.

6. S. Pinker, *The Language Instinct: How the Mind Creates Language* (New York: William Morrow and Company, Inc., 1994).

CHAPTER 16: MUSICAL NOTES

1. A. Aciman, "Are You Listening?," *New Yorker*, March 17, 2014.

2. D. Wright, *Deafness: An Autobiography* (New York: Harper Perennial, 1993).

3. E. Glennie, *Good Vibrations* (London: Hutchinson, 1990).

4. D. Bendor and X. Wang, "Cortical Representations of Pitch in Monkeys and Humans," *Current Opinion in Neurobiology* 16 (2006): 391–399; R. J. Zatorre, "Finding the Missing Fundamental," *Nature* 436 (2005): 1093–1094.

5. W. R. Drennan et al., "Clinical Evaluation of Music Perception, Appraisal and Experience in Cochlear Implant Users," *International Journal of Audiology* 54 (2015): 114–123; J. Schnupp, I. Nelken, and A. J. King, *Auditory Neuroscience: Making Sense of Sound* (Cambridge, MA: MIT Press, 2012); C. M. Sucher and H. J. McDermott, "Pitch Ranking of Complex Tones by Normally Hearing Subjects and Cochlear Implant Users," *Hearing Research* 230 (2007): 80–87.

6. K. Gfeller et al., "A Preliminary Report of Music-Based Training for Adult Cochlear Implant Users: Rationales and Development," *Cochlear Implants International* 16, no. S3 (2015): S22–S31.

7. B. Biderman, *Wired for Sound: A Journey into Hearing*, rev. ed. (Toronto: Journey into Hearing Press, 2016); M. Chorost, "My Bionic Quest for Bolero," *Wired*, November 1, 2005, https://www.wired.com/2005/11/bolero; Drennan et al., "Clinical Evaluation of Music Perception"; R. Wallace, *Hearing Beethoven: A Story of Musical Loss and Discovery* (Chicago: University of Chicago Press, 2019).

8. A. Romoff, *Hear Again: Back to Life with a Cochlear Implant* (New York: League for the Hard of Hearing, 1999).

CHAPTER 17: THE COCKTAIL PARTY PROBLEM

1. E. C. Cherry, "Some Experiments on the Recognition of Speech, with One and with Two Ears," *Journal of the Acoustical Society of America* 25 (1953): 975–979.

2. A. W. Bronkhorst, "The Cocktail-Party Problem Revisited: Early Processing and Selection of Multi-talker Speech," *Attention, Perception & Psychophysics* 77 (2015): 1465–1487.

3. J. O. O'Sullivan et al., "Hierarchical Encoding of Attended Auditory Objects in Multi-talker Speech Perception," *Neuron* 104 (2019): 1–15.

4. R. Litovsky et al., "Simultaneous Bilateral Cochlear Implantation in Adults: A Multicenter Clinical Study," *Ear and Hearing* 27 (2006): 714–731; R. J. M. Van Hoesel and R. S. Tyler, "Speech Perception, Localization, and Lateralization with Bilateral Cochlear Implants," *Journal of the Acoustical Society of America* 113 (2003): 1617–1630.

5. A. Romoff, *Listening Closely: A Journey to Bilateral Hearing* (Watertown, MA: Imagine Publishing, 2011).

6. J. Schnupp, I. Nelken, and A. J. King, *Auditory Neuroscience: Making Sense of Sound* (Cambridge, MA: MIT Press, 2012).

7. J. M. Hull, *Touching the Rock: An Experience of Blindness* (New York: Pantheon Books, 1990), 29–31.

8. Romoff, *Listening Closely*, 128.

9. S. R. Barry, *Fixing My Gaze: A Scientist's Journey into Seeing in Three Dimensions* (New York: Basic Books, 2009).

10. B. Crassini and J. Broerse, "Auditory-Visual Integration in Neonates: A Signal Detection Analysis," *Journal of Experimental Child Psychology* 29 (1980): 144–155; M. Wertheimer, "Psychomotor Coordination of Auditory and Visual Space at Birth," *Science* 134 (1961): 1962.

11. L. M. Romanski et al., "Dual Streams of Auditory Afferents Target Multiple Domains in the Primate Prefrontal Cortex," *Nature Neuroscience* 2 (1999): 131–136.

CONCLUSION: ATHLETES OF PERCEPTION

1. S. R. Barry, *Fixing My Gaze: A Scientist's Journey into Seeing in Three Dimensions* (New York: Basic Books, 2009).

2. Barry, *Fixing My Gaze*; O. Sacks, "Stereo Sue: Why Two Eyes Are Better than One," *New Yorker*, June 19, 2006; O. Sacks, "Stereo Sue," in *The Mind's Eye* (New York: Alfred A. Knopf, 2010).

3. W. James, *The Principles of Psychology* (New York: Henry Holt, 1890).

4. E. J. Gibson and A. D. Pick, *An Ecological Approach to Perceptual Learning and Development* (New York: Oxford University Press, 2000).

5. E. Goldberg, *Creativity: The Human Brain in the Age of Innovation* (New York: Oxford University Press, 2018).

6. Goldberg, *Creativity*.

7. See also P. F. MacNeilage, L. J. Rogers, and G. Vallortigara, "Origins of the Left and Right Brain," *Scientific American* 301 (2009): 60–67.

8. S.-J. Blakemore, *Inventing Ourselves: The Secret Life of the Teenage Brain* (New York: Public Affairs, 2018).

9. R. L. Gregory and J. G. Wallace, *Recovery from Early Blindness: A Case Study*, Monograph No. 2 (Cambridge, UK: Experimental Psychology Society, 1963), 33.

10. Blakemore, *Inventing Ourselves*.

11. S. E. Asch, "Effects of Group Pressure upon the Modification and Distortion of Judgments," in *Groups, Leadership and Men: Research in Human Relations*, ed. H. Guetzkow (Oxford, UK: Carnegie Press, 1951).

12. Blakemore, *Inventing Ourselves*; L. H. Somerville, "Searching for Signatures of Brain Maturity: What Are We Searching For?," *Neuron* 92 (2016): 1164–1167; A. W. Toga, P. M. Thompson, and E. R. Sowell, "Mapping Brain Maturation," *Trends in Neuroscience* 29 (2006): 148–159.

13. Toga, Thompson, and Sowell, "Mapping Brain Maturation."

14. E. M. Finney, I. Fine, and K. R. Dobkins, "Visual Stimuli Activate Auditory Cortex in the Deaf," *Nature Neuroscience* 4 (2001): 1171–1173.

15. M. Saenz et al., "Visual Motion Area MT+/V5 Responds to Auditory Motion in Human Sight-Recovery Subjects," *Journal of Neuroscience* 28 (2008): 5141–5148; H. Burton et al., "Adaptive Changes in Early and Late Blind: A fMRI Study of Braille Reading," *Journal of Neurophysiology* 87 (2002): 589–607; A. Pascual-Leone and R. Hamilton, "The Metamodal Organization of the Brain," *Progress in Brain Research* 134 (2001): 427–445; L. B. Merabet et al., "Rapid and Reversible Recruitment of Early Visual Cortex for Touch," *PLOS One* 3 (2008): e3046.

16. Pascual-Leone and Hamilton, "The Metamodal Organization of the Brain"; Merabet et al., "Rapid and Reversible Recruitment of Early Visual Cortex for Touch."

17. Saenz et al., "Visual Motion Area MT+/V5 Responds to Auditory Motion."

18. H. J. Neville et al., "Cerebral Organization for Language in Deaf and Hearing Subjects: Biological Constraints and Effects of Experience," *Proceedings of the National Academy of Sciences* 95 (1998): 922–929.

19. Saenz et al., "Visual Motion Area MT+/V5 Responds to Auditory Motion."

20. Barry, *Fixing My Gaze*.

21. Gregory and Wallace, *Recovery from Early Blindness*.

22. Chorost, *Rebuilt*, 171–172.

23. M. Ross, "A Retrospective Look at the Future of Aural Rehabilitation," *Journal of the Academy of Rehabilitative Audiology* 30 (1997): 11–28.

24. Ross, "A Retrospective Look at the Future of Aural Rehabilitation."

25. Barry, *Fixing My Gaze*.

26. Chorost, *Rebuilt,* 171.

27. B. Biderman, *Wired for Sound: A Journey into Hearing*, rev. ed. (Toronto: Journey into Hearing Press, 2016).

28. M. von Senden, *Space and Sight: The Perception of Space and Shape in the Congenitally Blind Before and After Operation* (Glencoe, IL: Free Press, 1960), 160.

29. Barry, *Fixing My Gaze*; Goldberg, *Creativity*; D. Bavelier et al., "Removing Brakes on Adult Brain Plasticity: From Molecular to Behavioral Interventions," *Journal of Neuroscience* 30 (2010): 14964–14971; C. D. Gilbert and W. Li, "Adult Visual Cortical Plasticity," *Neuron* 75 (2012): 250–264; E. Goldberg, *The Wisdom Paradox: How Your Mind Can Grow Stronger as Your Brain Grows Older* (New York: Gotham Books, 2005); A. Pascual-Leone et al., "The Plastic Human Brain Cortex," *Annual Review of Neuroscience* 28 (2005): 377–401; E. R. Kandel, *In Search of Memory: The Emergence of a New Science of Mind* (New York: W. W. Norton and Co., 2006); M. M. Merzenich, T. M. Van Vleet, and M. Nahum, "Brain Plasticity-Based Therapeutics," *Frontiers in Human Neuroscience* 8 (2014): doi, 10.3389/fnhum.2014.00385; Q. Gu, "Neuromodulatory Transmitter Systems in the Cortex and Their Role in Cortical Plasticity," *Neuroscience* 111 (2002): 815–835.

30. S. Anderson and N. Kraus, "Auditory Training: Evidence for Neural Plasticity in Older Adults," *Perspectives on Hearing and Hearing Disorders: Research and Research Diagnostics* 17 (2013): 37–57; D. M. Levi, D. C. Knill, and D. Bavelier, "Stereopsis and Amblyopia: A Mini-Review," *Vision Research* 28 (2015): 377–401.

31. A. Romoff, *Listening Closely: A Journey to Bilateral Hearing* (Watertown, MA: Imagine Publishing, 2011), 164.

32. Barry, *Fixing My Gaze*; Goldberg, *Creativity*; Bavelier et al., "Removing Brakes on Adult Brain Plasticity"; Gilbert and Li, "Adult Visual Cortical Plasticity"; Goldberg, *The Wisdom Paradox*; Pascual-Leone et al., "The Plastic Human Brain Cortex"; Kandel, *In Search of Memory*; Merzenich, Van Vleet, and Nahum, "Brain Plasticity-Based Therapeutics"; Gu, "Neuromodulatory Transmitter Systems in the Cortex"; Romoff, *Listening Closely*; P. R. Roelfsema, A. van Ooyen, and T. Watanabe, "Perceptual Learning Rules Based on Reinforcers and Attention," *Trends in Cognitive Sciences* 14 (2010): 64–71; S. Bao et al., "Progressive Degradation and Subsequent Refinement of Acoustic Representations in the Adult Auditory Cortex," *Journal of Neuroscience* 26 (2003): 10765–10775; A. S. Keuroghlian and E. I. Knudsen, "Adaptive Auditory Plasticity in Developing and Adult Animals," *Progress in Neurobiology* 82 (2007): 109–121.

33. Gregory and Wallace, *Recovery from Early Blindness*.

34. Biderman, *Wired for Sound*, 26–27.

35. Barry, *Fixing My Gaze*.

36. Boswell, *The Life of Samuel Johnson*.

37. Gibson and Pick, *An Ecological Approach to Perceptual Learning and Development*.

38. Bao et al., "Progressive Degradation and Subsequent Refinement of Acoustic Representations"; Keuroghlian and Knudsen, "Adaptive Auditory Plasticity in Developing and Adult Animals."

39. Chorost, *Rebuilt*, 126.

40. Romoff, *Listening Closely*; A. Romoff, *Hear Again: Back to Life with a Cochlear Implant* (New York: League for the Hard of Hearing, 1999).

41. Romoff, *Hear Again*, 159; Chorost, *Rebuilt*, 88.

42. Barry, *Fixing My Gaze*; Goldberg, *Creativity*; Bavelier et al., "Removing Brakes on Adult Brain Plasticity"; Gilbert and Li, "Adult Visual Cortical Plasticity"; Goldberg, *The Wisdom Paradox*; Pascual-Leone et al., "The Plastic Human Brain Cortex"; Kandel, *In Search of Memory*; Merzenich, Van Vleet, and Nahum, "Brain Plasticity-Based Therapeutics"; Gu, "Neuromodulatory Transmitter Systems in the Cortex"; Roelfsema, van Ooyen, and Watanabe, "Perceptual Learning Rules Based on Reinforcers and Attention"; E. R. Kandel, "Increased Attention to Spatial Context Increases Both Place Field Stability and Spatial Memory," *Neuron* 42 (2004): 283–295.

43. R. Latta, "Notes on a Case of Successful Operation for Congenital Cataract in an Adult," *British Journal of Psychology* 1 (1904): 135–150.

44. R. V. Hine, *Second Sight* (Berkeley: University of California Press, 1993), 82.

45. Bavelier et al., "Removing Brakes on Adult Brain Plasticity"; Gilbert and Li, "Adult Visual Cortical Plasticity"; Goldberg, *The Wisdom Paradox*; Pascual-Leone et al., "The Plastic Human Brain Cortex"; Kandel, *In Search of Memory*.

46. R. Kurson, *Crashing Through: A True Story of Risk, Adventure, and the Man Who Dared to See* (New York: Random House, 2007).

47. E. Huber et al., "A Lack of Experience-Dependent Plasticity After More Than a Decade of Recovered Sight," *Psychological Science* 26 (2015): 393–401.

48. Betty Edwards, *Drawing on the Right Side of the Brain: The Definitive*, 4th Edition (New York: Tarcher Perigree, 2012).

49. R. Feynman, *"Surely You're Joking, Mr. Feynman!"* (New York: W. W. Norton and Company, Inc., 1985).

50. E. J. Gibson and A. D. Pick, *An Ecological Approach to Perceptual Learning and Development*, 201.

Index

Rosalie Winard

Susan R. Barry is professor emeritus of biology and neuroscience at Mount Holyoke College, where she researched neuroplasticity and stereovision, and the author of *Fixing My Gaze: A Scientist's Journey into Seeing in Three Dimensions*. She lives in Massachusetts.